The 40-Knot Sailboat

The 40-Knot Sailboat

By Bernard Smith

E P B M

ECHO POINT BOOKS & MEDIA, LLC

ISBN: 978-1-62654-936-4

Published by Echo Point Books & Media
www.EchoPointBooks.com

Printed in the U. S. A.

To

the late DR. A. A. KNOWLTON,
Professor of Physics at Reed College,
who gave me my first useful insights
into the laws of nature.

ERRATA*

Page 92, column 1, lines 23-25.

These lines should read: that there is no substantial forward component to a sail's force until the relative wind is about 25 degrees off the boat's direction of motion.

Page 92, column 1, line 35.

Insert make substantial *before* progress

Page 96, column 2, lines 4 and 3 from foot.

For effectively doubled *read* greatly increased

Page 97, column 1, line 2.

For the doubled *read* an improved

Page 97, Figure 42, last line.

For is equivalent to *read* may be nearly

Page 107, Symbols.

Add to list:

K = foil geometry factor

For $K_5 = \dfrac{DK_1K_2H^3}{L}$ *read* $K_5 = \dfrac{DKK_1H^3}{L}$

Page 109, column 2, equation 18.

For $1/A_R = \dfrac{K_2(R^2-1/25)H^2}{(R-1/5)^2H^2}$ *read* $1\,A_R = \dfrac{K(R^2-1/25)}{(R-1/5)^2}$

Page 112, Figure 49.

Abscissa is "Offwind Angle," last value is 165 degrees. Ordinate is "Velocity" (Knots). Black curve for non-lifting aerohydrofoil; white for lifting version.

Page 117, column 2, line 17.

For 35·4 *read* 35 *and for* 13·5 *read* 13·4

Page 117, column 2, line 20.

For $V_b \ max = \dfrac{35\cdot4\ W^2}{13\cdot5^2}$ *read* $V_b \ max = \dfrac{35\ W^2}{13\cdot4}$

*The errata listed here are noted at the corresponding locations.

Contents

Illustrations

Foreword

Preparing this book on a new kind of sailboat for both technical and nontechnical audiences, became so complex a problem that the final decision to write the equivalent of three books came as a relief. The original version was tailored for the mathematically trained reader. Much of it has been retained intact in Part Three, which should lead the initiate from elementary principles, through some unavoidable mathematics, to the design criteria for the fastest sailboat in the world.

Part One tells the history of the sailboat's problems. Its inclusion was stimulated in part by the sheer weight of engrossing related material gathered in literature surveys, and in part by a frank recognition that the book as originally planned would reach only a restricted audience. However, to avoid a heavily labored book I have assumed from the start that the reader has at least a brushing familiarity with sailing practice.

A reader without a background in sailing will find it profitable to browse through the Appendix before reading the book. Readers with no knowledge whatsoever of fluid dynamics may have less trouble with Part Two (the History of the Aerohydrofoil) after perusing Chapters VII and VIII which are free of mathematical descriptions. The last four chapters of the book, however, do require at least a smattering of the physical sciences for comprehension. To aid the nontechnical reader as much as possible, the results of each technical analysis are stated in plain English and, where appropriate, have been brought forward into the narrative sections of the book.

I am indebted to Howard I. Chapelle, Curator, Division of Transportation, Smithsonian Institution; Col. John J. Wade, U.S.M.C.; Capt. Paul Van Leunen, Jr., U.S.N.; Framton Ellis, Bureau of Naval Weapons; and T. G. Lang, Naval Ordnance Test Station, for their kindness in examining the technical content for accuracy and adequacy. Howard I. Chapelle, in particular, has contributed ideas and counsel for which I am very grateful. I am also indebted to Charles E. Cutress, Associate Curator, Division of Marine Invertebrates, Smithsonian Institution, for his assistance in running down the interesting sidelights on *Physalia* and *Velella*.

John Newbauer, editor-in-chief, *Astronautics and Aerospace Engineering*, is to be especially identified as having made this book possible. The notion of writing the book and the preliminary arrangements essential to getting on with it resulted from his initiative.

The Disclosures

Since all men were children in their most impressionable years, it should come as no surprise to find that many of the strongest aspirations of grown men come not from experiences in adult life but from childhood dreams. My own fixation on the problem of the sailboat evolved from boyhood adventures in small boats. Through the accretions of time, this fixation gradually became decorated with sophisticated technical ornaments which were a natural, if indirect, product of adult training, and which could have been hung on other walls just as well.

One cannot remake history, yet I would guess that if at some critical time in my life the center of my attention had been captured by an object other than the sailboat, the nature of my experiences would have been much the same. I would be looking back in wonder—just as I do today, at the accidents of time that caused the many apparently unrelated sources of my knowledge to converge and to yield new insights into the problem of the sailboat—as if some purposeful destiny beyond my conscious control had directed these events.

I am reminded of the duck that now holds the limelight in scientific journals. At some critical time in its life the duck also exhibits a special readiness to admire an animated object. As soon as its eyes can see, the first moving thing the duck beholds becomes its mother. Ducklings have been caused to follow joyously moving footballs throughout their adolescence, merely because such an object was revealed to them at the proper time. Upon such tenuous threads hang the great differences between affection and disdain, and in this respect men may not differ basically from ducks.

The sailboat was the first inanimate object of my affection, and although I turned my attention to more remunerative interests, a subconscious corner of my mind remained devoted to the challenge of the wind. Even a quarter century of studied indifference to boating did not entirely put to rest an early compulsion to create the fastest sailboat in the world.

However, the passing years have had a telling effect on the quality of my dream. The signal

2

of achievement is for me no longer a transitory victory in a racing contest, but the solution of a difficult technical problem. The change may be likened to the situation of the composer who, drawn to his task originally through a love of listening to music, finds, after years of study, that he is perhaps more interested in the musical score than in the music. With the passage of time, he may lose his hearing, as some composers have, without losing the ability to write music. If I may stretch the analogy, the notes of sailing have become for me more fascinating than the sounds. Accordingly, this book is preoccupied with technical disclosures; not, however, to the exclusion of the many sailing values commonly appreciated by those who venture on the water's surface for profit or pleasure.

Wind as a Power Source

To ponder over the ethereal and elusive properties of the wind, that moody phenomenon that constantly reminds all witnesses of its limitless strength and subtlety, is a continuing source of excitement. Men have tried to chain it to useful purposes ever since the wind first demonstrated its catastrophic power, but they have not always succeeded. Wind-induced waves can hurl boulders weighing tons at massive masonry and in a twinkling smash it to rubble. To witness such destructive work on man's staunchest shore-based bulwarks is to conclude that the wind contains all the power man will ever need. But the wind is constant neither in magnitude nor in direction, and therein lies the problem of its mastery for man.

The wind is most freely available over that vast equipotential surface called the ocean—a surface having no hills or valleys except those of the waves, which are trivial and transitory compared to the irregularities of the land. The levelness of the ocean makes it an ideal surface on which to employ the wind to drive platforms, and men have always been quick to see this advantage. Far more of the wind's power has been used for propelling ships and boats than for all other purposes combined.

The energy of the wind does not have to be shipped or stored. It is the one source of power most always present where it can be used, and unlike sunlight, it can be tapped day or night, in clear or cloudy weather. But because it lacks the quality of constancy, man has been faced with a great dilemma. He has been unable to introduce into one simple machine the capacity to capture either a large part of a gentle breeze or a small piece of a gale, and therefore he has been unable to tap more than a fraction of the available power. Yet all the while he has been surrounded by more power than he needs.

Over the ocean, where the wind is most useful, the problem is especially complicated by the absence of a rigid base for holding a wind machine erect. Consequently, all attempts to drive sailboats at motorboat speeds usually end in disappointment. Either the boats capsize or unwanted properties are introduced with the increased speed. It is like dying of thirst while drowning. After centuries of struggle, the fastest sailboats of our time, whether clipper ships, America's Cup racers, inland lake scows or the amazing double-hulled canoes of the Pacific islanders, are, after all, only a little faster than the speediest vessels Magellan saw in his day.

In an art as old as sailing, any claim to a discovery of physical relationships that can double or triple the speed of racing boats may, understandably, be greeted with skepticism. Such doubt is not without good technical foundation. Over two hundred years of careful study have clearly shown that the price for small gains in the speed of keeled sailboats with good seakeeping qualities is an incommensurate increase in boat size, a penalty too great for the small purse of most sailing enthusiasts. Even a costly and elaborate hundred-foot yacht can hardly do better than twelve knots in a strong wind.

The Ancient 18-Knot Sailboats

Spared by a colossal ignorance of *what could not be done,* the Pacific islanders long ago invented smaller ocean-going sailboats capable of running at eighteen knots. The price paid for this unprecedented speed was the attainment of vertical stability only up to a certain wind strength and thereafter none at all. They knew nothing of the weighted keel and therefore turned to multiple-hulled boats, such as the catamaran and the proa in order to counteract the capsizing moment of the wind. In so doing they reduced the ballast requirement and inadvertently raised their craft higher in the water, thereby gaining a large profit in speed. Limited, however, by the state of their technical knowledge, they stopped short of the ultimate achievement. Not until comparatively modern times has there been so clear a recognition that the Pacific islanders were on the right track. To achieve significantly greater sailboat speeds the hull must, indeed, be raised higher and higher until at last it is lifted completely out of the water. The means for performing such a feat is the subject of these disclosures.

Definition of Aerohydrofoil

Stated in the simplest possible terms, the machine for deriving the ultimate in sailing speed consists of two vertical wings, an inverted one in the water joined to an erect one in the air. When coupled in this way the assembly may be likened to a sailboat that has a sail and a centerboard, but no hull; except that the sail is no longer a sail but an airfoil, and the centerboard no longer a centerboard but a hydrofoil. In fact it is no longer a traditional sailboat and therefore has been renamed an "aerohydrofoil."

However, nothing that runs over the water is a simple affair. As described above, the idealized aerohydrofoil could not possibly give a practical performance unless the problems of achieving buoyancy and stability without the aid of a hull were solved. It required the development of a new kind of hydrofoil that would combine the properties of a centerboard with buoyant support and dynamic lift, all within the same surfaces. It required the invention of an airfoil that would combine the flexibility of fabric sheets with the efficiency of a wing. And it required a unique arrangement of airfoils

and hydrofoils to obtain stabilizing forces that were independent of ballast or the strength of the wind.

These properties have been demonstrated in part with scale models and in part have been derived from theory. When all the evidence is accumulated, the prospect of attaining sailboat speeds considerably greater than twice the speed of the wind on a reach, and speeds equal to the wind when beating or running, becomes a real one.

So cursory a disclosure of the features and potentialities of the aerohydrofoil is bound to leave a disarming impression of its simplicity. In truth, the aerohydrofoil has few parts, but each performs simultaneously a large number of complex functions, and each function is inextricably entangled with all the others. A number of these functions have their origin in dim prehistory. Others have been added in recorded times. And some very crucial ones have been introduced in recent years—highly refined functions that make the aerohydrofoil at all possible.

I know of no better way of presenting these matters than through a historical recounting of the classical sailboat problems. This I have done, more for my own sake than for the reader's. When I undertook to describe the properties of the aerohydrofoil I had no thought of going back into time as far as I have, but curiosity triumphed. In the process I have learned more about how men came to be as they are. These findings I gladly pass along to the reader as, first, a study in one aspect of man's many-sided culture; and second, a way of chronologically introducing the successive functions added to the sailboat. I must emphasize, however, that the chronology is not presumed to be complete, but is only a recounting of critical technical highlights. The reader should look elsewhere in the literature for more detailed scholarly histories of the sailboat.

Problem of Man-Carrying Model

Why hasn't a man-carrying aerohydrofoil been built and sailed? Because its subtleties are more delicate even than those of a soaring glider. Its design poses problems in all of the six degrees of freedom inherent in an airplane. It must undergo the same elaborate development applied to an advanced airplane, and more—since it operates in two fluids, the wetter one of which can behave like concrete when struck at 40 knots. A machine moving at such speeds must be stable and controllable or it is dangerous. Although there exist modern tools of analysis and experiment to help make the first man-carrying aerohydrofoil successful, these tools are rarely if ever sufficient. And just as design errors on occasion are revealed in an airplane after it is airborne, so may they be expected in the aerohydrofoil after it is waterborne.

Pseudo-aerohydrofoils have been constructed in the past, large enough to be manned; some with hydrofoils and some with airfoils but none with both, so far as I can learn, save for my own scale models. These early large-scale versions, including one of my own, demonstrated in one way or another the advantages of aerohydrofoil principles as well as the shortcomings of incomplete aerohydrofoils. Now at last, the theoretical groundwork and the model testing are completed and the design of full-scale, true aerohydrofoils can be undertaken with optimism.

I have decided not to wait for the full realization of my dream before disclosing the new principles, knowing full well that if I must continue to work alone, some years will pass before the practical aerohydrofoil is developed. My purpose in making the disclosures at this point is to do all in my power to avoid a repetition of the time lost between the designs of da Vinci and those of the Wright brothers; and, by promulgating the theory and the understanding, I hope perhaps to bring other and better minds to the problem of marrying the theory, proved feasible in models, to full-scale practicability.

A Lonely contest with The Wind

"But why," asked a close friend of mine, "do you keep puttering with the aerohydrofoil? If it's speed you want why don't you get yourself an outboard motor?" I could feel the foundations of our friendship shaking before the question was completed, but fortunately the ties that bind us were strong enough to stand the strain. There is no direct answer to this sort of question. The exhilaration of sailing cannot easily be explained even to the best of friends; it can only be shared.

To one who has turned lifeless materials into a thing alive and forced them to do his bidding against the resisting forces of nature, in silence, without fuel, and without defiling the air or the water, there can never be anything more wonderful than the sailboat. To one who has not had the experience, no telling of it can touch him: The sailboat offends the senses of neither fish, fowl, nor man. To make it go faster is to make it even more a thing of freedom and beauty.

From the moment I tried, like so many other grammar-school boys before me, to sail a crudely shaped shingle and a square patch of cotton cloth on the Shrewsbury River in New Jersey, I felt drawn to a marvelous contest that surpassed any other I had ever known. The adversaries are human beings on the one hand, and the undesigning forces of a natural environment on the other.

Unless one stoops to overemphasize the more primitive aspect of this contest exemplified in sailboat racing, it need not be a zero-sum game like cards, chess, tennis, or football, wherein one contestant wins only on condition that some other contestant loses. Sailing is not a sport like hunting or bullfighting that requires the unprovoked death of some animal before a victory can be declared; yet once engaged, the contest can be just as vital—to the point where one's own life may be at stake.

Sailing, like mountain-climbing, is a contest that permits a man, alone if he chooses, the rare privilege of matching wits with an adversary who never wins or loses. Only the man wins or loses. And, as in mountain-climbing, high skill and refinement of taste must be developed to play the game or the player may not always know when he has won or lost. For the adversary never tells him by word or expression, and when the game becomes dangerous, no spectators, no umpires and no cheerleaders are on the site to bolster his morale, or to laud his performance.

This is what sailing has meant to me as I have moved from my first toy boat to timorous adolescent ventures across little pieces of the Atlantic Ocean in a small sailing canoe; to catboats, sloops, and iceboats in my early twenties; to sailing a 45-foot cutter on the Pacific as a young man; to a renewal of interest in midlife, stimulated, strangely enough, by the dry deserts of California and strengthened more recently by exhilarating experiences on Narragansett Bay.

The contest still favors the wind too much. Spurious tricks that the wind can play will, I fear, never be completely parried by better weather forecasting. A necessary part of an effective countermeasure is greater speed. All too frequently the sailor caught at the edge of a rapidly brewing dark squall, sometimes in dangerous currents, must frantically douse all sails and, with helpless anxiety await a drubbing. To retreat with pride and dignity, one

should be capable of riding the front of the rapidly moving disturbance well away from its heart. He who has ignorantly tried it with an ordinary boat, as I have, and has found himself either capsized, or with mast, sail, and rigging carried overboard in a tangled mess by the overtaking storm center, can never forget the experience. A boat which can attain a speed greater than that of the wind, like an iceboat, would bring some badly needed equity into the situation.

Under such conditions, thermal power is not of much help to the sailors of small pleasure craft. Motors have a nasty habit of conking out or of not starting when there is real trouble afoot. And even the largest private motorboat cannot carry enough fuel to run before a storm all day long at 30 knots. What is required is a speed equal to or greater than the wind, regardless of the wind's direction. Then, and only then, will the means to cope with the peril be proportional to its source. The aerohydrofoil may yet develop such sustained speeds over an oceanful of trouble without, in the process, frightening all the fish.

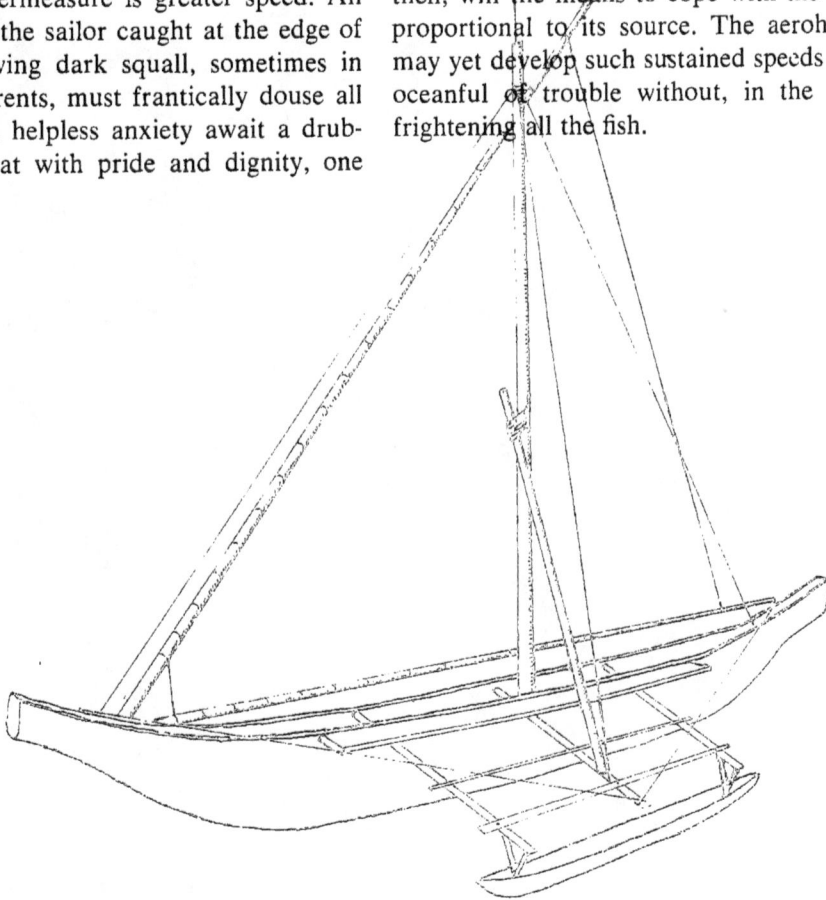

PART ONE

History of the Sailboat Problem

FIGURE 1
The Portuguese Man-of-War
(*Physalia physalis*)

The first sailboat to ride the ocean was some ancestor of the colorful Portuguese Man-of-War. The flat, flexible extension of the gas-filled float is employed by the jellyfish as a crude airfoil. This remarkable adaptation has helped the animal to control its position and to avoid being blown about at random by the wind.

CHAPTER I

Origins of the Sailboat

For practically every one of mankind's free-moving platforms the animal kingdom has an ancient counterpart. The most nearly perfect examples of nature's anticipation are the fish and the birds, both of which still surpass what men can do. No submarine can yet jump out of the water and glide over it like a flying fish; no airplane can yet dive into the water and swim like a gannet. Such exploits provoke the wonder of all who witness them. Yet, to me, the most remarkable of them all is the animal predecessor of the sailboat.

Nature seems to have abhorred the surface of the ocean when she marked out the various habitats for her creatures. Not many kinds of animals live solely on the surface of the sea. A few mollusks, an insect, and certain jellyfish seem to have this zone pretty much to themselves. The rest employ this interface for merely transitory purposes. But, from among the few true surface animals, one is the counterpart of the sailboat.

On the wavecrests of the far-flung oceans rides *Physalia physalis,* a beautiful purple jellyfish, romantically known as the Portuguese Man-of-War (Figure 1). It derives its common name from a historical identification with the coasts of Portugal and a fancied likeness to sailing warships of the early days. This gorgeously colored animal may well have sailed the seas long before the birds flew and fish swam, for it belongs to a phylum of animals far more primitive than the vertebrates. Below the waterline it is much like other jellyfish, dragging the usual tentacles and stinging cells with which it captures small prey. Above the waterline *Physalia* displays a remarkable organ: a float in the form of a sail. The float is a streamlined, gas-filled membrane, tapering upward to a sharp, high crest. Muscles attached to various parts of the float can change the shape of the sail, erecting or lowering the crest, bending the ends of the float to the left or right, or lowering them into the water.

It would be absurd to endow *Physalia* with a conscious desire to use the wind for propulsion. It does not have the highly developed nervous system that accompanies such motives.

However, a school of such animals sailing on a reach, all in consort, all moving at the same pace and in the same direction is enough to convince the observer that there is purpose to this behavior.

According to some marine biologists, early in life each of these animals is prompted to sail on one tack, perhaps nevermore to come about on the other. Whether or not it does come about is not known, since the event of changing tacks has never been observed. However, when the wind changes *Physalia* has been observed to turn its float and to sail on the same angle off the wind. Some have been known to curve their sails like an airfoil to match the direction of the wind, thereby gaining aerodynamic efficiency.

A Zoological Controversy

Not all students of the Man-of-War agree with this description of its habits, however. Dr. Charles E. Lane, Professor of Marine Sciences at the University of Miami, in a letter to me, expressed doubt that the *Physalia* found off Florida does little more than drift with the wind and current. On the other hand, A. H. Woodcock, of the Woods Hole Oceanographic Institute, who studied *Physalia* in the waters about Hawaii, witnessed animals sailing on port and starboard tacks that differed in direction by about 90 degrees.

Woodcock surmised from the data he collected that in the northern hemisphere *Physalia* is predominantly a port-tacker. This point of sailing is favored less and less as southern latitudes are approached. Such habits correspond to the reversal of wind and current circulation in the two hemispheres, but the relationship of these facts to the life cycle, if any, remains to be traced.

When not feeding, *Physalia* retracts its tentacles, thereby reducing the total wetted surface and total resistance to motion. I do not know whether or not the retracted tentacles are selectively bunched to increase leeway resistance, but I would strongly recommend that this possibility be investigated. If this phenomenon

should be observed it would, in my opinion, fully support the contention that *Physalia* is not an accidental sailer.

A smaller cousin of *Physalia,* the blue *Velella,* exhibits similar habits according to Manfred Curry, author of *Yacht Racing.* Marine biologists I have consulted do not believe that the anatomy of the animal permits such behavior. I have carefully examined detailed drawings of *Velella* and I must conclude that I see no organs for controlling a course of sailing in this animal. Nevertheless, I am sure that if the wind should accidentally strike *Velella's* sail in the right direction, *Velella* would tack! Not having the advantages of the Man-of-War's size, however, *Velella* is more easily subject to the buffeting of the elements; and, after a storm, one can frequently see the delicate cartilage-like skeletons heaped on the shore. As a boy I often observed them on the New Jersey coast, without knowing that these bodies were the remnants of sad shipwrecks at sea.

The Russians, who seem not to hedge on any matter, also have had something to say of late on *Physalia* and *Velella.* A. I. Savilov of the Institute of Oceanology, Academy of Sciences, U.S.S.R., unequivocally attributes the distribution of these two members of the order Siphonophora to their port and starboard sailing abilities. (Perhaps this matter is yet another in which the Soviets hope to gain a propaganda advantage by taking a firm stand while the Western world vacillates.)

Dawn of an Idea

Through an evolutionary process enduring millions of years, these jellyfish have learned to mitigate some of the wind's force in waters that probably were also the scene of man's earliest sailboat experiments. Only man has been able to duplicate the debated exploits of the Portuguese Man-of-War; and his first thoughts of sailboats may well have been stimulated by the sight of this lowly collection of organized matter, sailing accidentally perhaps, but appearing to sail nonetheless.

Regardless of how the idea originated, the

sailboat was the first machine to give men freedom of motion without muscle power. Few of us any longer recognize that the sailboat was truly the first instrument which freed us from bondage to the land. This place in history does not belong to the first log, reed float, or inflated skin on which man drifted downstream, nor to the first proper boat he paddled on a lake, nor even to the first ship driven by the wind on the sea, but to the first sailboat that could move upwind. This invention made the whole world accessible to man, a species of land animal, who suddenly found that an ancient barrier to mobility, the sea, could be converted into an excellent highway.

Before The Rocketship— The Sailboat

Neither do we recall, unless our attention is forcibly drawn to it, that the sailboat was the first machine to achieve powered motion without rotating parts. We have forgotten because we have been allowed little time to remember. Pelted and titillated daily by the advocates of modern science and industry through every channel of communication, our attention has now been diverted to the rocketship, which is also a marvelous machine requiring no rotating parts for locomotion.

Both machines—one many thousands of years old, the other younger than our lifetimes —have opened worlds at once vaster, richer, crueler, more traversable, and more mysterious than the land. But the wonder of modern technology that makes outer space accessible is not more admirable or inspiring than the enormous insight and courage of the first men who placed their confidence in bits of wood and matting, cleverly contrived to shackle the wind in the very teeth of the wind, and thereby journeyed to lands far beyond the reach of their muscle.

As one who has sought to understand the subtleties of both sailboats and spaceships, my greater esteem goes to our ancestors who, long, long ago without benefit of scientific theory, brought the sailboat to a high state of refinement. Civilized man has taken seven hundred years to progress from the "fire-arrow" of the Chinese to the rocketship. Primitive man might have taken as long as seven thousand years to go from the log to the sailboat. Considering his handicaps, his was the greater accomplishment.

The Anomaly of Upwind Sailing

The first thoughts of "catching the wind" with mats or sheets to push a boat come early and easily. But the notion of driving boats upwind *by using the wind itself* seems to be contrary to everyday experience. It seems to deny common sense. Few men have ever had the courage to defy the dictates of common sense which, after all, is nothing more nor less than what everybody believes at a particular time in history. It might have been easier for the ancients, who had successfully domesticated animals, to think of harnessing porpoises or fish for the purpose of sailing upwind. Indeed, there are old legends that may have had their origins in such thoughts.

We need not view our forebears from lofty pinnacles of superiority; well-educated people, trained in science and engineering, have often astonished me by asking how a sailboat can go against the wind. They were actually laboring under the delusion that a sailboat can travel only where the wind blows it and, therefore, that it must await the right wind to move in the desired direction. Usually a fair amount of "unlearning" must take place before an understanding of the true situation can be reached. Many modern men with technical backgrounds and with vector diagrams at their disposal, come hard by the notion of beating to windward. Far less apparent were such notions to early man.

The most striking evidence we have of the elusiveness of this concept is the total absence of such boats among the Eskimos or Indians of North America before its discovery. These were among the most resourceful people on the face of the earth, people who had developed the kayak and the canoe into forms so highly refined that they are copied to this very day. Nor

is there a word in any report by the first chron-
iclers to indicate that any Aztec or Mayan had
a recognizable knowledge of sailing before the
white man arrived. The curious fact is that
there were millions of people in North and
Central America who had long coastlines and
a need for coastal trade, who enjoyed a civilization often likened to that of the biblical Egyptians; who had an astronomy and a numbering
system and other arts superior to the Egyptians,
but who apparently never sallied forth in sailboats as did the Egyptians. This situation is especially remarkable when we are reminded of
the visits of the Norsemen to the New World.
Both the Eskimos of Greenland and the Indians
of New England must have been exposed to the
sight of the Norsemen's sailing ships many centuries before Columbus, yet neither of these
people were stimulated to improvise a sailboat
—perhaps because they did not comprehend its
properties, or felt no need for it.

The Disputed Drop-Keel Raft

Only in South America is there any evidence
that Indians practiced sailing of sorts. Strangely
enough the alleged locales were on opposite
sides of the continent, not far from the Equator;
one in ancient Peru, the home of the Inca, and
the other thousands of miles away in Brazil at
the site of present day Recife. Coupled with this
is the odd coincidence in the design of the craft
at these two points. Both were rafts and both
were reputed to carry drop keels placed between the logs, which presumably gave them an
ability to tack.

I must confess that when I am confronted
by this much coincidence my suspicions are
aroused. Very large gaps exist in the history
of these two craft, the "balsa" of Peru, and the
"zangada" of Brazil. Shortly after the first
sketchy reports, during the conquest of South
America by Europeans, the use of sailing rafts
·with 'drop keels remains unnoticed, or at least
unmentioned, for very long periods of time.
This could easily have been the result of the
cruel acts of persecution practiced by the white
man who virtually stamped out entire Indian

cultures in the Americas. Even so, I am still
compelled to wonder whether such sophisticated craft were pre-Columbian in the first
place.

The log or reed raft is so much older than
true boats that time may have permitted independent sailing experiments with this ancient
platform in different parts of the world. The
sailing raft may very well have been the common precursor of the single and the multiple-
hulled sailboat, but hardly in a form that was
capable of tacking.

The great problem is not the presence of
sailing rafts in such widely separated locales
as Recife, the Peruvian coast and Formosa, but
the simultaneous use of the highly sophisticated
drop keel by people who hardly could have
been in communication with each other. This
very clever and valuable sailing device was
invented by Europeans not before the late 17th
century, and then only in connection with true
hulls. We find well-documented evidence of the
independent and earlier invention of only one
similar device—the variable depth rudder of
the Chinese Junk, which undoubtedly existed
at the time of Marco Polo. And once again it
appears to have been applied first to the true
hull.

All around the coasts of the China Seas very
resourceful modifications of the droppable junk
rudder have been employed for centuries to
diminish the side-drift of sailboats. The For-
mosan raft sailors operating in the same seas
could, quite logically, have come to their drop
keels through a process of borrowing and mod-
ifying.

In each case, for Chinese and Europeans,
the drop keel arrived on the scene after these
highly civilized maritime cultures had spent
centuries developing the true sailing ship. The
pre-Columbian Indian on the other hand was
a traditional landlubber. If he did have the drop
keel, it was the result of a very rare accident,
not in keeping with his cultural evolution either
before or after the Spanish invasion. It hardly
makes any difference today as to whether the
story is a legend or not; it had no determining
influence on his destiny.

12

Origin of The Tacking Sailboat

True tacking sailboats were probably invented only twice in the history of the world; over 4,500 years ago along some coastline not far from the Middle East, and somewhere in the Pacific at an unknown later date. There is a great and fundamental difference in the craft coming from the two areas—so great that the independence of their origin is all but certain. The Middle Eastern version, which is associated with continents, is always single-hulled. The Pacific type, which seems to be indigenous to small islands, is always multiple-hulled.

No early record of any civilization along the entire Afro-Eurasian coastline mentions any multiple-hulled sailing boat, be it catamaran or outrigger. Conversely, no single-hulled sailing boats were ever reported by the first Europeans who explored the outlying Pacific Islands, over the whole region identified as Oceania, excepting possibly New Zealand.

In order to remain erect under the side force of the wind, the boat that appears to have originated all around the coastlines of the Arabian peninsula depended on keeping the center of gravity inside a single hull. From this corner of the globe the spread of similar boats most probably moved westward throughout the length of the Mediterranean Sea to Western Europe and concurrently eastward into the Indian Ocean, possibly continuing around Malaya to China.

The Pacific Islands boats were kept from overturning by providing a very broad buoyancy base, most effectively accomplished by spacing two or more hulls as widely as possible. Precisely where in the Pacific this idea was conceived is unknown. By the time it came to the attention of Europeans it had spread all over the Pacific, but without reaching the Americas or eastern continental Asia, or Japan, although it had entered the Indian Ocean, probably via the Indonesian chain of islands.

Trend Away from Symmetry

There exist a few consistent trends in boat development that may provide a basis upon which the course of sailboat history can be traced. The most primitive floating platforms invariably are the most symmetrical—in fact the earliest, the log, has the highest order of symmetry. It has no front, back, left, right, top or bottom, except as determined by the way a man sits or stands on it. Because a log has radial symmetry it has no roll stability whatsoever, requiring, therefore, the combined talents of a navigator and acrobat to paddle or sail it.

A raft has less perfect symmetry and, consequently, more stability. Its top and bottom cannot easily reverse position, solving the first and most important problem of any boat. In the process of improving this particular stability characteristic, all subsequent boats have been designed with above-water contours that are distinctly different from those below the waterline.

The first proper boats almost always had identical sides and identical ends. As the functions of the boat became more specialized, more asymmetry was introduced. To improve performance either bow and stern were made dissimilar or left and right sides lost interchangeability. The fastest single-hulled sailboats had the highest asymmetry of the first kind; the fastest double-hulled sailboats the highest asymmetry of the second kind.

A Hypothetical History of Multiple-Hulled Boats

The history of the single-hulled boat is fairly well documented. But the story of the multiple-hulled boat has many gaps. To aid in the process of bridging the gaps, a number of hypotheses, that can be established with reference to the known history of the single-hulled boat, may be applied to the multiple-hulled version.

The most general hypothesis is perhaps valid for all of man's cultural pursuits. Long usage of an implement or a mechanical aid. or long practice of a particular art, tends to retard the introduction of improved tools or practices. Once a contrivance becomes part and parcel of a particular culture it is not easily displaced by better contrivances.

14

The second hypothesis involves the increasing asymmetry associated with more advanced boat hulls, which has already been discussed. The same tendency can be discerned in the development of the sail, which has progressed from the symmetrical square sail to the more efficient asymmetrical fore-and-aft sail.

The third hypothesis is more specialized insofar as it stipulates the conditions for the development of single-hulled boats. Single-hulled boats are associated with continents where the incidence of lumber forests and navigable rivers are common; multiple-hulled boats with islands, where suitable boat building lumber is limited and where boat development in relatively calm river water is not likely.

Taken together, the three hypotheses spell out the following history for each kind of boat.

A. SINGLE-HULLED SAILBOAT

Originated on navigable rivers; migration followed continental coastlines close to lumber forests and harbors. Place of origin contains sailboats with clearest vestiges of fore-and-aft symmetry in current sailboats; greatest differentiation between bow and stern found at points of most recent introduction, which coincide with locales having most highly refined sailboats.

B. MULTIPLE-HULLED SAILBOAT

Originated on island beaches; migrated to other islands but not to continents where it could not displace single-hulled boats. Place of origin has sailboats with greatest bilateral symmetry; greatest bilateral asymmetry and refinement found at points of farthest migration and most recent introduction.

Verification of this hypothesis would necessitate long and laborious research for which there is little place in this book. Let me sum up my inferences at this point by drawing parallels between Western and Eastern sailboat evolution that may indicate the migratory path of the Pacific boat culture.

Certainly the North Atlantic coasts were the last regions to improve the single-hulled sailboat and there also evolved vessels with the most highly refined fore-and-aft asymmetry. Authoritative historical evidence points to the Middle East as the indigenous source of the single-hulled sailboat and in that area it is still almost as primitive in symmetry as it ever was. The path from one region to the other can be traced with reasonable assurance.

The hypothesis would also support a contention favoring a similar migration eastward as far as the China Seas. Not only do we find many of the stages leading from the Arabian dhow to the Chinese junk along the eastern route but the direction of evolution matches the path of migration. The greater sophistication and asymmetry is on the eastern end.

The remarkable parallelism at the extremities of the two possible migrations strongly suggests a fixed pattern for evolutionary development once the single-hulled sailing ship was born. Not until the 15th century, and only sporadically for several centuries thereafter, were ship and boat builders on European and Chinese seacoasts in communication with each other. Yet both cultures eventually and independently contrived through radically different designs, the rudder, the fore-and-aft "control" sail and the drop keel. Figure 2 plots the courses conjectured for the eastward and westward expansions.

In the case of the Pacific Islands and the double-hulled boats, we can identify at present only a localization of the evolutionary end, which seems to have taken place in Micronesia. The northernmost extension of this grouping, namely the Marianas, a tongue of islands reaching toward Japan, was the home of the fastest and most highly asymmetrical sailboat in all the Pacific. This was the proa, a double-hulled canoe with a large main hull and a small outrigger attached to the windward side. The more symmetrical catamaran was prevalent farther east. It seems to have had earlier origins, for its geographical range is contiguous, which is not true for the proa. This evidence suggests that the expansion of the multiple-hulled sailing canoe was probably westward in its later stages.

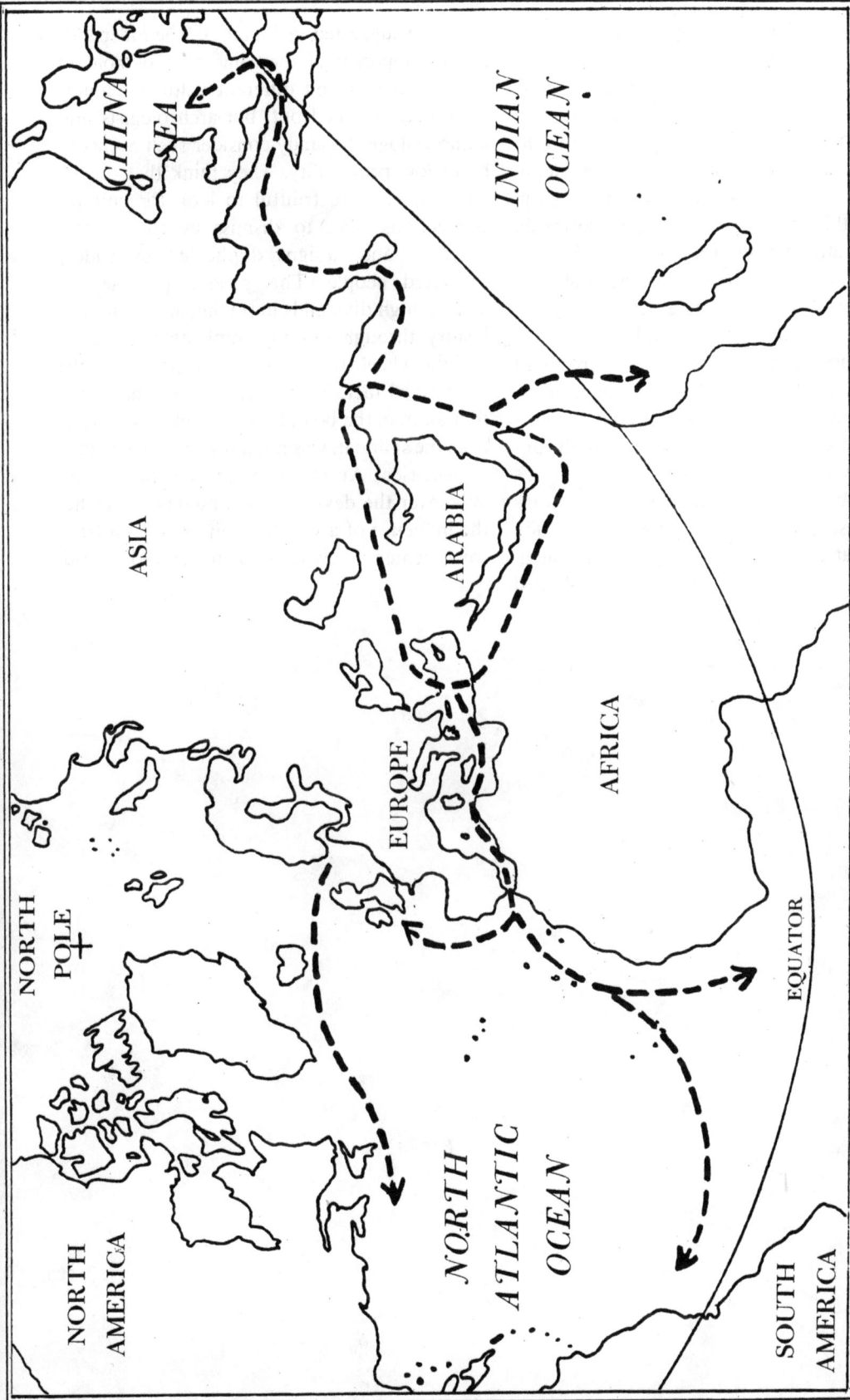

(*Oblique Mercator Projection by the author.*)

FIGURE 2

Migration of the Single-Hulled Boat

The area within the marked triangle is the most probable home of the first single-hulled sailboats. The western routes of migration, shown by the dashed arrows, are generally verified in historical records. The eastern route to the China Seas is suggested by some similarities between the boats found along the route and older boats formerly found in the marked triangle.

From Raft to Catamaran

The evolution of a Polynesian double canoe from a raft is not difficult to imagine; one can think of hollowing out the logs of a raft to make it lighter for beaching and of reducing the number of logs to two to further reduce the weight. If the people who came to Polynesia arrived in rafts, it is entirely possible, that the succeeding craft they built with the materials at hand, were not at first single canoes but multiple-hulled dugouts. This would help to explain the predominance of the multiple hull throughout the Pacific. Perhaps it is no accident that "catamaran," which today is used in reference to the double canoe, was originally applied to the wooden raft.

It must be admitted that these observations are at best superficial and require further study before great credence can be placed in them. The migratory paths of the multiple-hulled boats are suggested in Figure 3. The independent development of the same kind of boat in different places and different cultures cannot, of course, be ruled out, but archeologists and anthropologists usually consider such events to be of low probability. They think that on the whole it is more fruitful to look for cultural connections than to dismiss as coincidental practices and designs duplicated by widely separated people. This general premise has yielded high dividends in tracing man's cultural history through hunting implements, pottery, and the wheel. In support of this premise is the established fact that the wheel, more necessary to man than the boat, and conceptually simpler than the sailboat, was not indigenously invented in America, Australia, or the Pacific Islands. Wherever the device is found today it signifies the influence of a visitor or an immigrant from some center of civilization in the Old World.

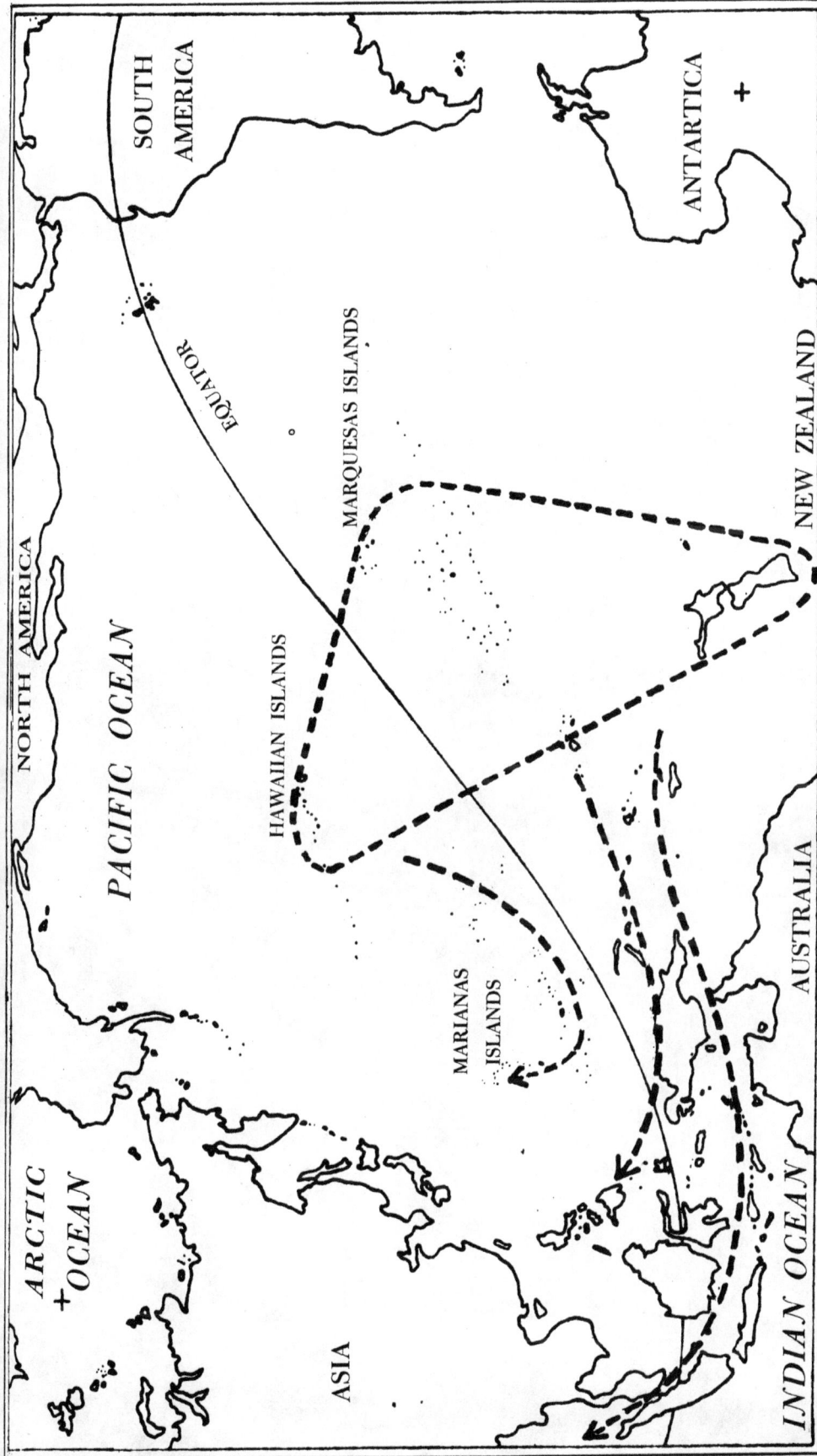

(Oblique Mercator Projection by the author.)

FIGURE 3

Migration of the Multiple-Hulled Boat

The general culture, as well as the common use of the catamaran, mark the triangular area bounded by the Hawaiian Islands, the Marquesas Islands and New Zealand as one in which the natives most probably had common ancestors. The Oblique Mercator Projection largely preserves the relative distances and directions in the entire Pacific Basin and shows the enormous expanses the Polynesians would have had to traverse to reach the triangle. The proa is found more commonly to the westward of the triangle, and the fastest of the proas, along the route marked by the northernmost arrow.

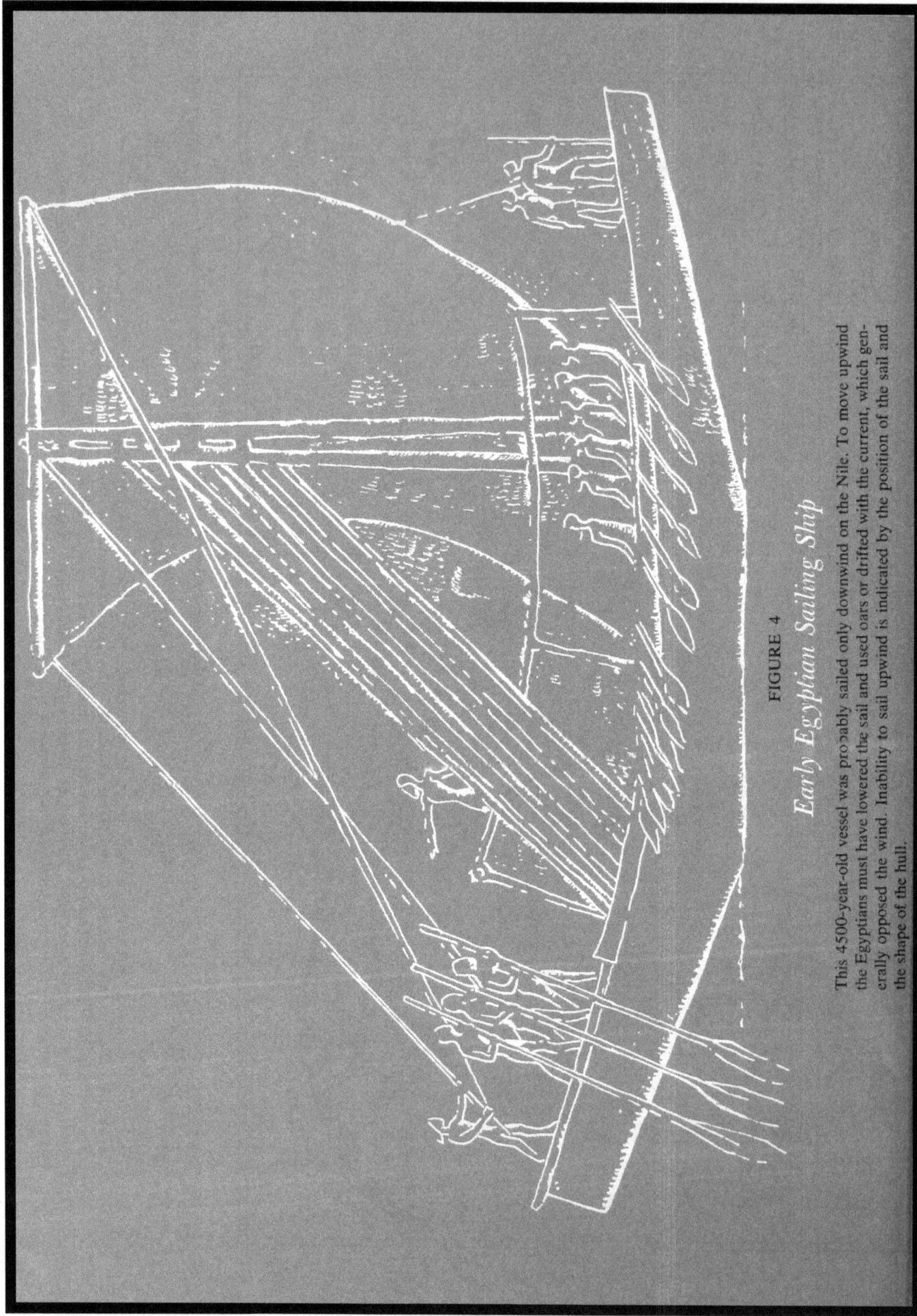

FIGURE 4

Early Egyptian Sailing Ship

This 4500-year-old vessel was probably sailed only downwind on the Nile. To move upwind the Egyptians must have lowered the sail and used oars or drifted with the current, which generally opposed the wind. Inability to sail upwind is indicated by the position of the sail and the shape of the hull.

CHAPTER II

Ancient Approaches to the Solution

The earliest records of sailing vessels come from the Egyptians and are about 4,500 years old. A stone carving from that age is reproduced in Figure 4. So much detail has been depicted that there is no difficulty in concluding that such a ship could hardly beat to windward. The shape of the hull was not destined to resist leeway, nor is anything else in evidence that could prevent drift except the oars, and it is doubtful that they were used for this purpose at that time. The forward position of the mast and sail also militate against sailing to windward. With the sail this far forward, the strength of the steersmen would be taxed to keep the boat on anything but a downwind course in a strong wind.

In all probability it was a river boat that drifted or was rowed down the Nile with its sail down, returning on the wind that generally blows upstream in the Nile valley. Nile boats are operated exactly the same way today. Of interest is the bipod mast, which eliminated the need for shrouds. It may surpirse some sailboat designers, who occasionally come up with the same idea, to learn that it is thousands of years old.

Five hundred years later, Cretan artists pictured boats like those of the Egyptians, except that they had a cutwater. As illustrated in Figure 5, a cutwater is a flat, forward extension of the keel and prow. Probably the combination of cutwater, forward sail, and strong men on the steering oars produced the first ship that had a decent chance of sailing upwind. The great Phoenician navigators must have had ships of the same kind, which possibly explains why they were willing to make extended journeys. They knew they could return, even with the wind against them.

The cutwater probably was derived from the forward ram of early Mediterranean warships. So long as it remained, either as a vestige of the ram or as the ram itself, it contributed to better windward performance. We do not have a very clear picture of the merchantmen of those days because most archaeological sources depict only the more glamorous fighting ships, which invariably carried a ram as well as oars-

FIGURE 5

Early Cretan Sailing Ship

(*From a reconstruction by Björn Landström*)

Four thousand years ago, shipbuilders in the Mediterranean Sea had apparently discovered a crude way to give a vessel some upwind sailing qualities. By extending a flat prow to form a cutwater, the side drift on a tack was diminished. The Phoenicians adopted the same device, which probably enabled them to make long voyages without oarsmen.

men. Quite possibly many kinds of merchant-men in the Mediterranean Sea possessed good sailing qualities because the underwater ram was retained.

Shortly after the beginning of the Christian Era, the cutwater disappeared from most of the merchant-ship representations found on coins or stones of the time. This may not signify the disappearance of this feature; it may simply mean that it became less important. Moreover, it is not always possible to determine what the underwater lines of a ship are like when an ancient illustration shows it floating with highly stylized waves all about it. It is quite probable that the need for the cutwater diminished with the development of deeper- and bulkier-keeled vessels, though it must have been retained on smaller boats. Some form of the cutwater is used on many small, shallow-water boats all over the world today.

During several periods of history, attempts were made to expand the advantages of the cutwater by extending a similar surface at the stern, which, in my opinion, indicates a clear recognition of how it aided the windward working of a boat. However, without the rudder post and the true rudder, the stern cutwater (or skeg) must have been more of a hindrance than a help. It certainly made the steersman's job more difficult, since he had to overcome the directive force of the skeg with the smaller surface of an oar located in a position of poorer mechanical advantage. Such boats must have been sluggish in their response to steering efforts.

Roman Merchantman

About the second century A.D., there appeared decked Roman merchantmen that were forerunners of many European, African, and Asiatic ships for a thousand years thereafter. Their masts, yards, rigging, bowsprits, main-sails, topsails, foresails, deck structures, and anchors heralded the pattern of ships at least up to the Age of Discovery. Figure 6 illustrates how modern they really were. There was no provision for rowing, which is good reason to

believe that they had windward sailing capabilities. These ships seem to have appeared on the scene full blown. However, the consensus is that their development took many centuries; in one form or another similar ships must have existed long before they were recorded in stone or bronze. In view of the large territory covered by the Roman Empire in that period, this design may very well have influenced the pattern of ships all over the civilized world. No one knows how far its influence extended, but it is interesting to note that some Japanese junks have the same fundamental form even today.

Lionel Casson, author of *The Ancient Mariners*, has uncovered ancient Greek stones, dating from this period, that show ships similar to the Roman merchantman but fitted with fore-and-aft sails. Some are clearly sprit sails and one is unquestionably a forerunner of the lateen sail. Casson has presented strong evidence for Greek knowledge of fore-and-aft sails as early as the second century B.C.

Evolution of Lateen Sail

After the third century A.D. there seems to have been little basic change in sailing ships until the thirteenth century, except for the increased use of the *Latin* sail, "lateen" sail, as it was called by northern Europeans. The refined version of this sail was probably an invention of the Arabs (Figure 7). Its use spread throughout the Mediterranean region and the Indian Ocean, just as if it had followed the same migratory path that is conjectured for the first single-hulled sailing ships.

The evolution of the lateen sail is not difficult to imagine. It may have been nothing more than a tilted square sail at first. Indeed, some of the old pictures seem to show the square sail being used this way, and some modern Philippine boats hold their square sails exactly in such an orientation. I do not wish to attribute to the ancient mariners a knowledge of aerodynamics, but I think we can grant them the sagacity to recognize that by tilting the square sail they could reach stronger and steadier winds aloft without increasing the size of the mast or yard. Regardless of how it developed, the lateen sail is a real aerodynamic improvement over the square sail for windward sailing. Small wonder that pirates all around the Mediterranean basin and along the northwestern waters of the Indian Ocean seized upon this rig to give them an advantage in running out to meet the booty, which could not escape by turning upwind.

The clearest pictures of the lateen sail come from the Greeks of the ninth century, and it holds a firm place in subsequent nautical history. Its ascendency was so rapid that records hardly mention the square sail for several hundred years thereafter. In all probability the square sail did not disappear, but the Mediterranean peoples, concentrating on refining the lateen sail, were not likely to have been especially interested in any contemporaneous square-rigged ships.

We know that northern European seamen used square sails exclusively until well past the Middle Ages, perhaps in ignorance of what was going on farther south. The northern boats could hardly be considered improvements over the Cretan boats developed three thousand years earlier.

Viking Ships—
Behind the Times

Into this class fall the famed ships of the Norsemen, who may have trusted more to their muscles than to their sails (Figure 8). Perhaps ignorant of the true sailing art, the early Vikings may have hoisted sail only when the wind was with them. At other times they probably strained their backs at the oars of their large whaleboats, which were designed more for rowing ease than sailing. There is some evidence that in the tenth century they did have boats with shallow keels suitable for moderate beating to windward. Whether or not such boats were used to reach America can only be guessed.

Howard I. Chapelle, Curator, Division of Transportation, Smithsonian Institution, who has studied the question for many years, believes the Norsemen came to America in boats

that did have some ability to work to windward. I think he reasons correctly when he suggests they probably were "too lazy" to row across an ocean against the prevailing westerlies. My own experience supports his contention. After adding a sail to a small canoe I once owned, I never paddled it again regardless of whether there were winds or calms. Even when I lost a leeboard one day, I persisted in making snail-like progress upwind rather than resort to paddles—I preferred arriving home late to saving time at the cost of brute work.

Seamen of ensuing centuries gradually developed elaborate combinations of square and lateen sails. The emphasis had shifted back to the square sail by the fourteenth century, and this sail remained in the ascendancy until the mercantile use of the sailing ship declined in the nineteenth century.

Advantages of True Rudder

The true rudder also stems from the fourteenth century, which suggests a possible connection with the square sail's ascendancy. As a matter of fact, many of the maneuvering difficulties that plagued the early square-rigged vessels, when moving on a tack, could have been overcome by a well-designed rudder. The very tail end of a ship is, after all, the best place to put a control surface if one must counteract a heavy, unbalanced wind force. A rudder also has a clear engineering advantage over a steering oar because it is attached to the hull precisely where the control force is generated—below the water line. A steering oar, on the other hand, must not only sustain the torsion generated in steering a boat, but also the bending moment produced by supporting the oar at some distance above the blade. Of course the ability to sweep an oar and swing a boat's stern rapidly, even though the boat moves slowly, is lost with a rudder, but this advantage is confined to relatively small boats. The steering oars of the large ancient ships were too heavy to be handled as sweeps and were mounted in fixed positions to relieve the steersman of excessive fatigue.

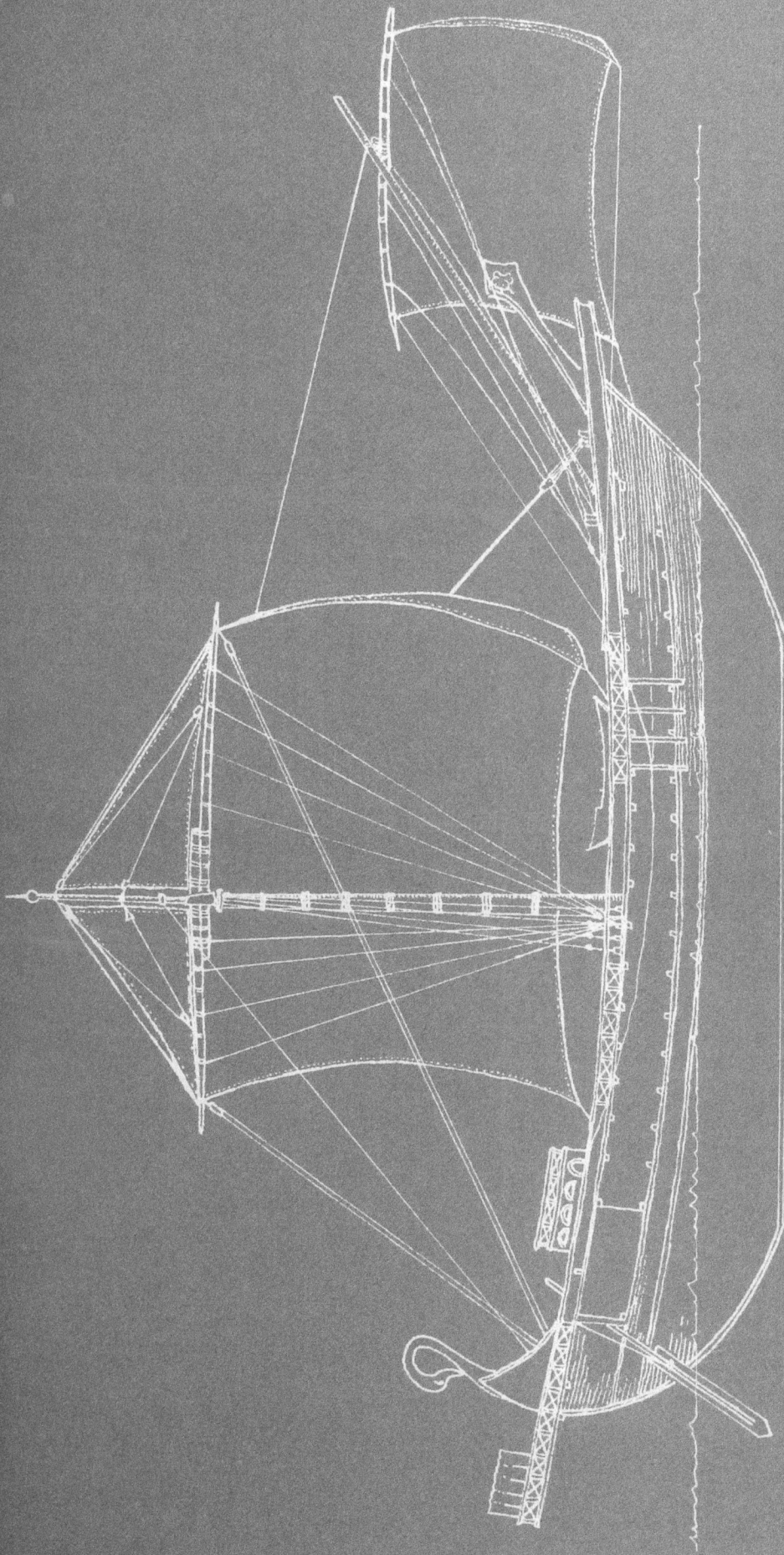

FIGURE 6

Roman Merchantman

(From a reconstruction by Björn Landström)

Early in the Christian Era ships of this design were common trade carriers in the Mediterranean and other parts of the world. They were deep and bulky, and were sufficiently good sailers to dispense with oarsmen. The general form of this ship and its sails set a basic pattern that lasted a thousand years.

FIGURE 7

Lateen Sail

(*From a reconstruction by Björn Landström*)

For beating upwind the lateen sail, introduced 2,000 years ago and refined in the Middle Ages, was decidedly superior to the square sail. Without increasing the size of the mast or the yard, the lateen sail reached higher to intercept stronger winds and it curved to a better aerodynamic shape.

FIGURE 8

Viking Ship

The famed ships of the Norsemen of the ninth century were, in general, inferior to contemporaneous Mediterranean ships. Their sailing qualities were no better than those of vessels used three thousand years earlier in the Mediterranean Sea. Without deck, true keel, lateen sail, or protection for the steering oarsman, the Viking ships were far behind their times. The illustration portrays a "long ship," which was the Viking warship. The "round ship," a cargo carrier, may have been a better sailer. Some round ships of this period may have been furnished with a deck, as well as with differentiated bow and stern like those used in cargo carriers of later centuries.

FIGURE 9

Carrack

By the fifteenth century there existed a fairly complete understanding of ship's rigging. The carrack, fitted with a square foresail and mainsail, a lateen mizzen, a rudder, and a compass, was ready for the Age of Discovery. Columbus' *Santa Maria* was a carrack. A similar ship of the time, the caravel, was fitted with lateen sails only to improve its windward sailing qualities.

The rudder probably allowed maritime traders to make better use of the increasing wind knowledge that was being accumulated during these times. Knowing when and where prevailing winds were available, a trader could plot a course to coincide with downwind sailing, in which case a square sail would be more effective than a lateen sail. Moreover, sailing downwind required less ballast, movable or fixed, and less ballast always signified more payload to a trader. His built-in rudder could help him run on a tack when his planning was poor or if nature were unco-operative.

The lateen sail never really vanished on the larger sailing ships, but it began to serve a new purpose. More and more it became a *control* surface, like the tail of an airplane. By the time of Columbus there was fairly complete understanding of how to hold a ship on course by trimming a lateen sail near the stern (Figure 9).

In the small-boat field the lateen sail has held its own up to modern times. (As a matter of fact, the first sail I made for my first sailboat was a lateen sail.) It probably had an independent origin in the Pacific, where it was invariably used with a boom as well as a yard. The boom was not extensively used in Europe until much later.

Sailing Ships in The Ancient Far East

We have little to guide us in discerning where the rest of the world stood during these early years. Marco Polo's description of Chinese ships, along with later information, indicates that the Chinese were ahead of the Europeans in the thirteenth century. By this date they had large ships fitted with genuine rudders and multiple sails arranged for good control to windward (Figure 10). Not only do the Chinese appear to have understood the control function of sails, but they may also have made progress in using the backwash of forward sails to increase the "lift" of adjacent sails, much as a modern jib does for a mainsail. There is little question that they knew of watertight bulkheading to increase the safety of ships, and that they had magnetic compasses. With all this they were certainly equipped for longer sea journeys than were the Europeans, and they actually did venture into the Indian and South Pacific Oceans. Yet they made no attempt to exploit their discoveries. One explanation may be that they were too busy developing an inland empire to be bothered with maritime expansion, much as the Russians have been until modern times.

No one knows for sure how the Chinese came to their lug sails. It is possible to imagine a progression leading to this design by starting with a square sail set to a yard and boom. The second step could have been a combination square and lateen sail. Such a sail could not be brought from one side of the mast to the other. The Chinese could have observed a loss of driving power when the wind came from the side that caused the sail to form a double curve around the mast. Stiffeners, in the form of additional horizontal spars, could then have come as a natural solution. The system of multiple flexible stiffeners and multiple sheet lines attached to the stiffeners gave the Chinese very good control of the sail's camber and twist throughout the sail's height. In this way the entire sail could be adjusted and curved for upwind as well as downwind sailing.

The junk's rudder was most ingenious. Not only was it designed to be raised and lowered like a modern centerboard, but in some junks it was placed so far forward that to some extent it also performed the *function* of a centerboard. Through centuries of trial and error the Chinese found a combination of control sail and rudder which, in basic principle, was not different from the control system improvised for the aerohydrofoil. Moreover, they knew precisely how to take much of the manual strain from the manipulation of large sails and

FIGURE 10

Chinese Junk

About two centuries before the Europeans created the carrack or the caravel, the Chinese had developed a superior ship that in all essentials remains unchanged to this day. The junks of Marco Polo's time had more efficient hulls, rudders, and sails than European ships of the same period. These old Chinese ships had many modern features, such as watertight bulkheading and "control" sails.

FIGURE 11

Clipper Ship

The clipper ship of the 19th century was designed primarily for sailing close-hauled, although it made its best speed on broad reaches. To compete in the growing Asiatic trade contest it had to contend with winds, such as the monsoons, that seasonably opposed its course. The highest speed recorded, credited to the *James Baines*, was 21 knots. The clipper, of course, was a merchantman and not a racing ship, which makes even more remarkable the comparison of its speed to those of displacement racing sailboats.

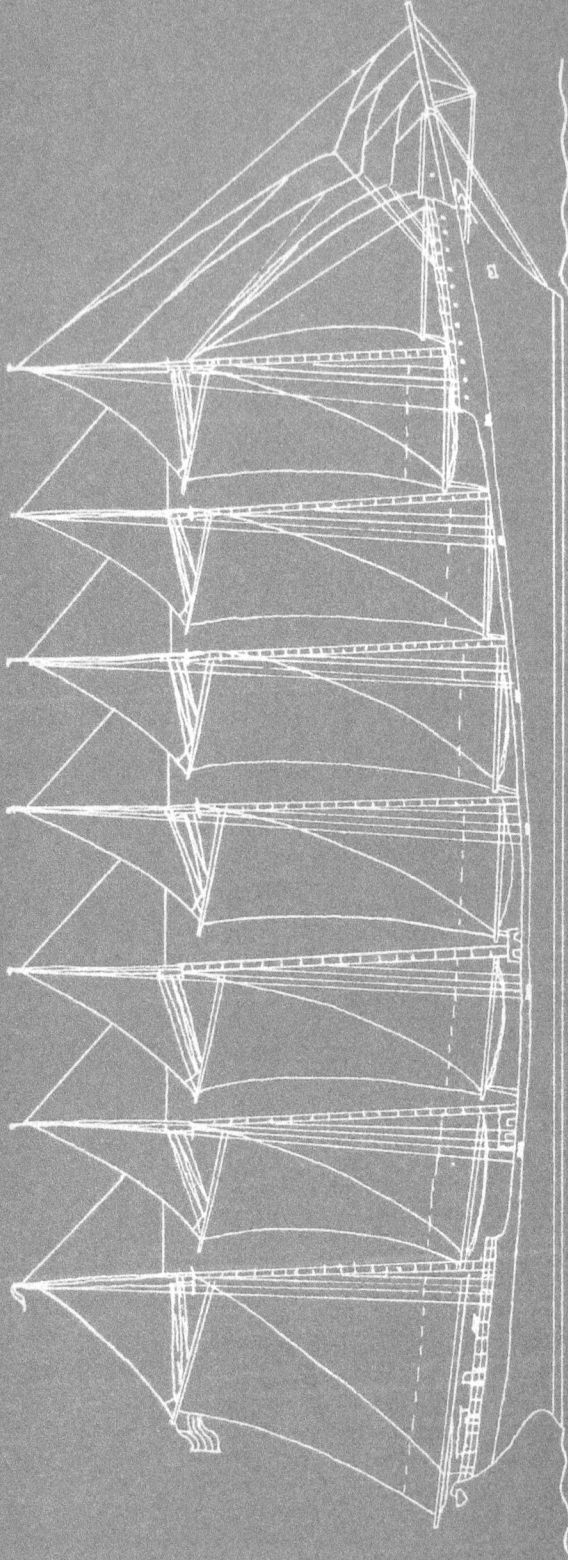

FIGURE 12

Commercial Schooner

The seven-masted schooner was an extreme form of the sailing ship in the final competition with steam. Fore and aft sails were employed exclusively because their setting and trimming required less manpower than was needed for square sails. Despite the use of such sails, these schooners were not especially fast or particularly efficient at beating to windward. The airflow over each sail successively changed the direction of the wind, making it necessary to set the after sails at high angles to the true wind. The forces developed by the after sails partially cancelled those developed by the forward sails.

rudders by adding balancing areas in the right places. Undoubtedly this was discovered first in the Chinese lug sail, where, by adding area to the sail *ahead* of the mast, the stresses on the sheet lines could be reduced as desired. By the same process of adding area ahead of the hinge line, the Chinese were able to diminish the force required to control the rudder.

The Maritime Explosion

By the sixteenth century there was common knowledge of all fundamental sailing ship principles in Europe, the Far East, and many places between. The ensuing maritime explosion can be attributed solely to European ambitions, with the rest of the world as its object. In the contest for empire the emerging victor, England, reached its pinnacle through the instrument of the sailing ship. English sailors were the first to learn the true circulation of the world's winds, and with this advantage they used the roaring forties in the Southern Hemisphere for free transport to all parts of the world.

It would be going too far afield to give even a cursory listing of all the sailing ship mutants over the next two centuries. The variations seem to be more decorative than functional and are more representative of rococo art than of technical improvement.

The most spectacular developments were in matters of size and complexity, occurring mostly around the North Atlantic basin and building up to a climax in the extravagant clipper ships of the nineteenth century (Figure 11). The fastest of such ships, the *James Baines,* attained a speed of 21 knots. But this record and others close to it were made on broad reaches before gales, with clouds of canvas straining every spar and mast almost to the breaking point. Clipper ships could make no such speed beating to windward. Although the clipper ship was designed to sail close-hauled better than any prior square-rigger,

except the North Atlantic packet-ship, its most spectacular speed was attained when the wind was abaft the beam. In this respect it represented the culmination of a 5,000-year effort to reach downwind at high speeds. It had come to the end of the line, and it could go no further because the nature of the vessel itself stood as a barrier.

On the other hand, while the larger ships were being pushed before the wind, smaller ones were being developed for better upwind sailing. The subtleties and refinements of these less spectacular accomplishments deserve separate treatment. They are best discussed in the context of modern sailing problems, even though some of the approaches are traceable to the sixteenth century.

Last Commercial Sailing Ships

In the final phases of commercial sailing a convulsive effort was made to compete with steam by turning to fore and aft sails. The steamship was winning, not so much through greater independence of weather and routes, but primarily because of lower labor costs. Sailing ship operators hoped to regain the economic advantage by converting to fore and aft sails (Figure 12), for which fewer hands were required. Despite the lower operating costs and greater versatility displayed by the newer sailing ships, however, they were not able to overtake the steamship, which through additional technological improvements had gained an unbeatable advantage.

The history of the commercial windjammer is an eloquent lesson on the illusion of cheap or free power sources—a lesson we seem to forget rather easily. The same naïve expectations have been held out for sun power, tide power, wave power, thermal power in ocean currents, and, today, nuclear power. The hard, cold realities are that the factors of capital investment and operating costs frequently overrule the cost of the energy source.

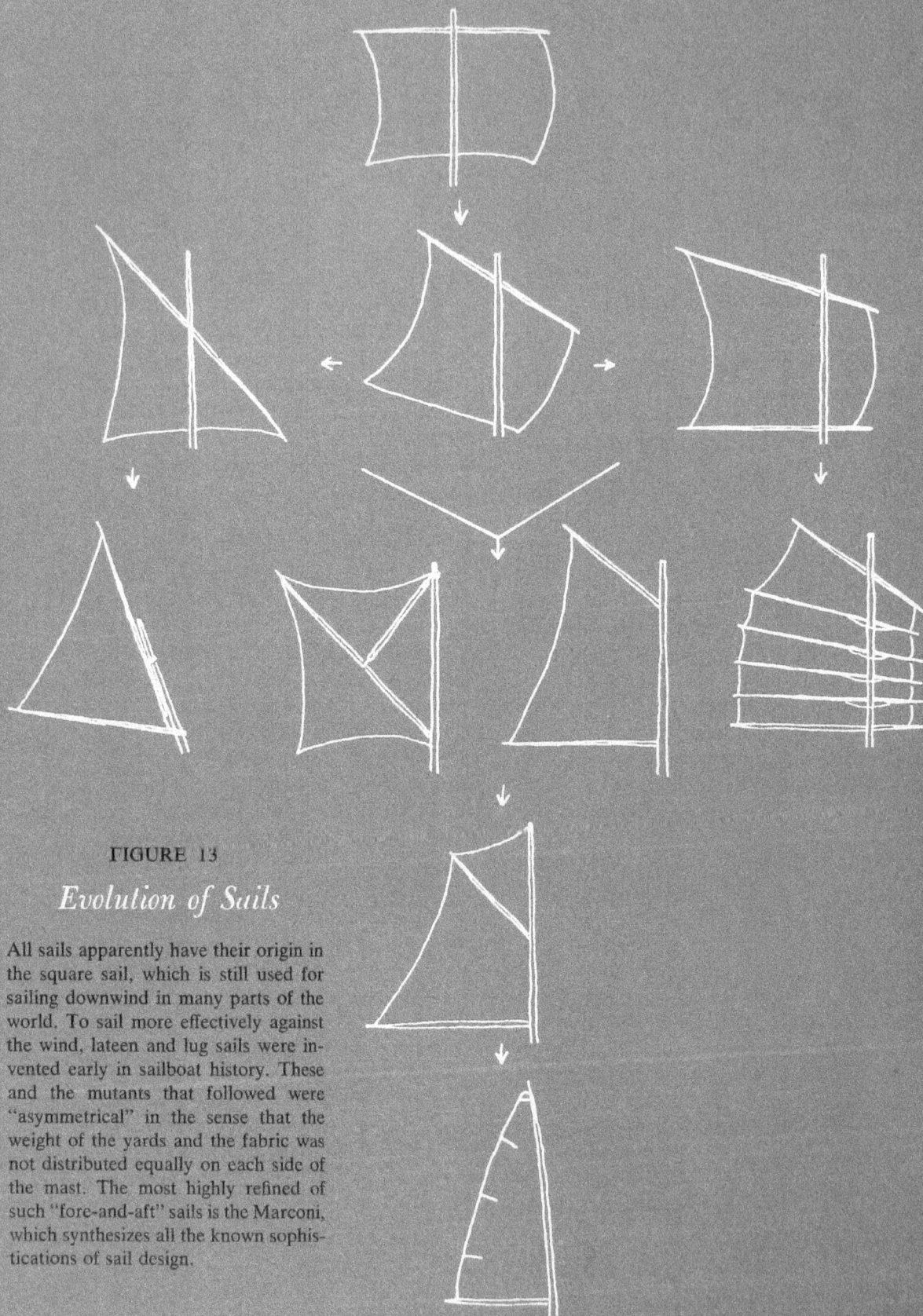

FIGURE 13

Evolution of Sails

All sails apparently have their origin in the square sail, which is still used for sailing downwind in many parts of the world. To sail more effectively against the wind, lateen and lug sails were invented early in sailboat history. These and the mutants that followed were "asymmetrical" in the sense that the weight of the yards and the fabric was not distributed equally on each side of the mast. The most highly refined of such "fore-and-aft" sails is the Marconi, which synthesizes all the known sophistications of sail design.

CHAPTER III

Modern Approaches to the Solution

In the seventeenth century there was a great flowering of new sail shapes. Concurrently, more attention than ever before was given to underwater lines. Within the next two centuries small boats were developed that incorporated new features in almost all possible combinations, from which were derived today's pleasure and racing sailboats.

These efforts were quite apart from the gross approaches of warship and merchant-ship designers. In many respects the methods of attacking the problems could be called "modern." The sails became more like airplane wings, and, for that matter, so did the devices projected into the water to resist leeway. Attention was being devoted at long last to making surfaces more efficient, rather than larger. Without delving into the chronology of the developments, the evolution of the two kinds of devices can be seen in Figures 13, 14, and 15.

In Figure 13 we trace the popular Marconi sail from its progenitor, the square sail. (The Marconi was so named because the rigging resembles a wireless tower.) It is possible to trace several paths, any or all of which lead to the Marconi. Not all sailors concur in the belief that the Marconi is the best of sails. A surprisingly small amount of systematic wind-tunnel testing has been conducted to demonstrate its superiority, and very little of what has been done has been published. However, the Marconi does have a pleasing shape, and if good form follows good function, as claimed by the best industrial designers, then the Marconi should be the paragon of sails.

The development of underwater shapes leading to the most efficient fixed keels (Figure 14) has apparently followed only one path. On the other hand, the moveable appendages generally known as drop keels underwent a number of widely divergent evolutions (Figure 15). Two in particular are of special interest because they illustrate how separate

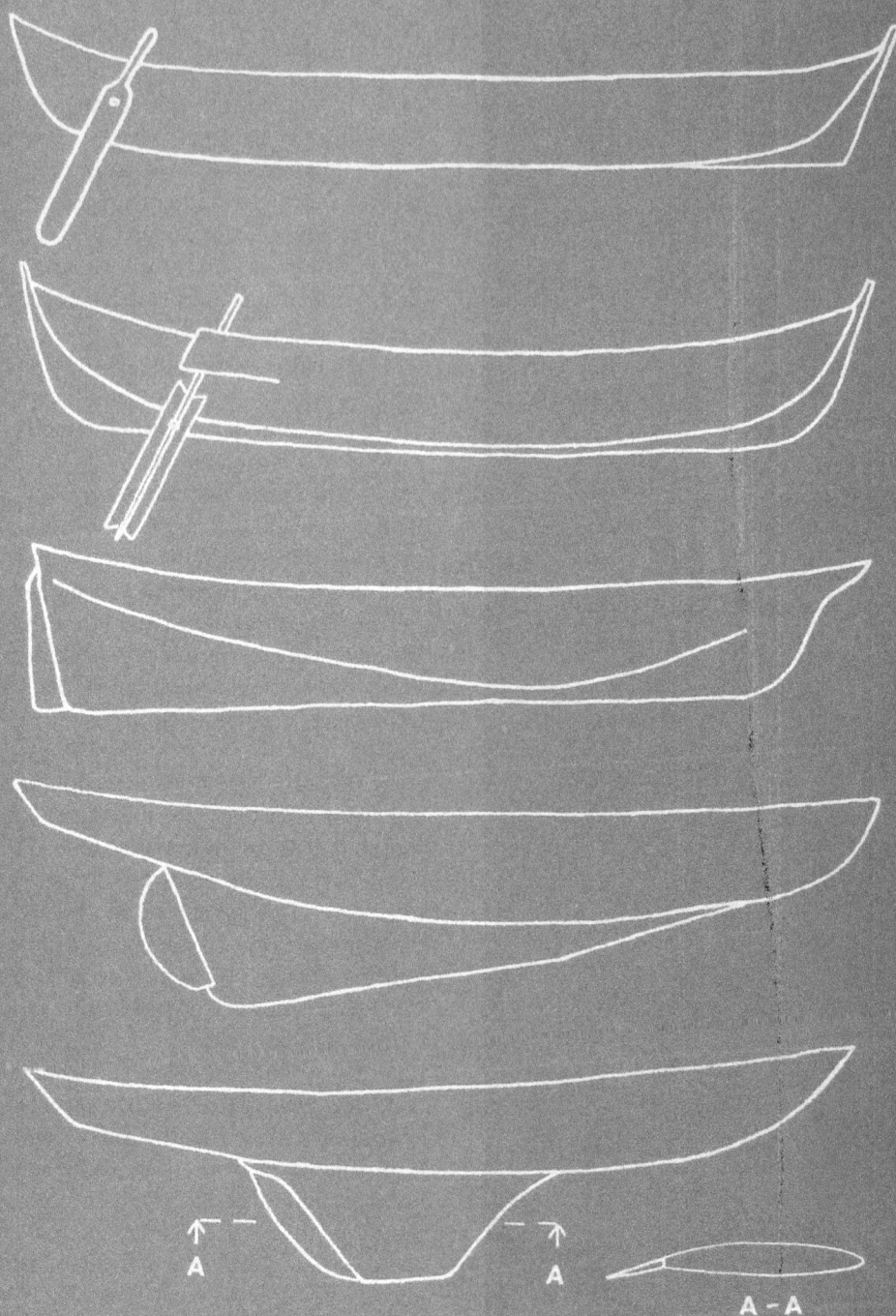

FIGURE 14

Evolution of Sailing Hulls

The vertical underwater surface, needed to counteract the wind's side force, seems to have been introduced at the very front of sailboats. From that position it migrated backward, to be concentrated most heavily near the stern. In modern times, however, it has been moved forward, closer to the hull's center. In the process of its migration the sailboat keel developed lines that closely resemble an airfoil, as shown in section A-A. Turning the rudder to neutralize weather helm effectively gives the keel an angle of attack, which is present in other arrangements only when the boat sideslips.

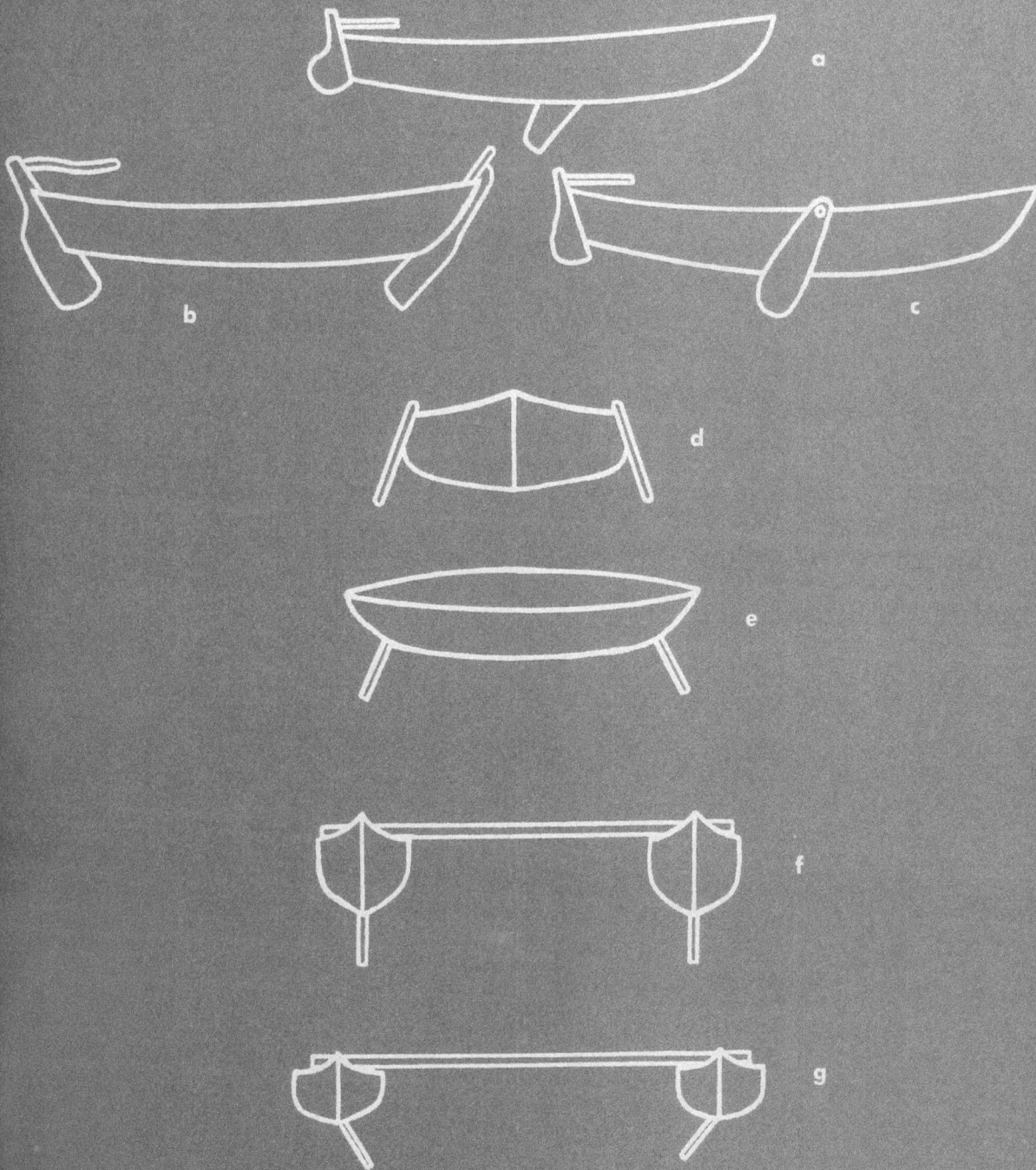

FIGURE 15
Movable Devices for Resisting Leeway

Articulating keels, or drop-keels, are conveniences which were invented primarily for shallow-water sailing. They also provide a fringe benefit insofar as they can be raised to reduce drag when sailing downwind. Three symmetrical drop-keel arrangements are shown in sketches (a), (b), and (c). These are, respectively, the common centerboard, the Indo-Chinese rudder and sternboard, and the Dutch leeboards. The front views are (d) leeboards and (e) scow bilge boards, both of which are inclined to allow for heeling; (f) double centerboards for a catamaran; and (g) double centerboards for a sailplane, inclined to produce righting moments under side forces. (See Figure 23)

FIGURE 16

Sailboats and Modern Analogues

The modern yawl or ketch is aerodynamically equivalent to half an airplane furnished with a slot wing. The forward control surface used in some missiles and proposed for some futuristic aircraft (also used in early aircraft) was employed in sailboats hundreds of years ago. To some extent, a jib is used on many boats today for the same purpose.

cultures sometimes solve exactly the same problem, yet the solutions can be identified readily as independent. Aside from breaking into the hull of a boat with a centerboard, there exist only two symmetrical external arrangements for drop keels: at the sides of a hull and at the bow and stern. The Dutch chose the first form and invented swiveling leeboards, but, oddly enough, the Indo-Chinese chose the more elegant second form and merely added a single stemboard to their junks. Thus, with a deep rudder and a deep stemboard, both of which were fitted to slide in slots, the Indo-Chinese solved the problem of leeway resistance in shallow waters with one board less than the Dutch.

When sails are properly positioned and set, they make the modern racing combination of keel and rudder especially efficient, as shown in Figure 14. This advantage is not available in ordinary centerboard boats, but the same effect and more is potentially available to leeboard boats and to boats with separate drop keels for each tack, like the scow and catamaran. Without making leeboards more complex, their efficiency can be improved greatly by curving each one to fit the corresponding tack, and by setting each at an angle that would prevent any drift of the boat whatsoever. Some lake scows have bilge boards arranged in this fashion. The Dutch, who developed leeboards, shaped them independently for each tack long before the hydrodynamic theory was formulated. The Dutch even made provision for changing the angle of attack in accordance with sailing points and wind strengths. These changes are made automatically as a leeboard rotates about its pivot point, its movement controlled by the shape of the guard on which it bears.

Cultural Lags

In Figure 16 two arrangements of fore-and-aft sails are displayed to reveal some analogies with more modern inventions. Few aerodynamicists know that they have only re-discovered what sailors invented hundreds of years ago;

nor do many sailors know that millions of years ago the birds beat them to every trick—including the slot effect created by the jib and mainsail.

The most astonishing demonstration of a cultural lag is shown in Figure 17. Here we see how the lateen sail, known for over a thousand years, is employed to create a "new" flying device. This wing, formed by joining the feet of two lateen sails, has been called everything but a sail, almost as if the connection were not apparent. It is interesting to note that sailmakers have at last been invited to advise on the fabrication of the paraglider. In recent models the fabric runs perpendicular to the leech of the wing, which is the correct method. No modern sailmaker would sew fabric strips parallel to the leech, as shown in Figure 17. Further, good sailmakers could show the paraglider designers how to increase sail area with minimum added weight by incorporating a "roach" and battens into the trailing edge. These features would also improve the shape and increase the lift-to-drag ratio.

If one searched through history to find examples of man's distressing inability to use his highly-developed brain for transferring knowledge from one discipline to another, he could find no better illustrations than those given above.

Refinements in Single-Hulled Racing Sailboats

The refinements of combined keel and rudder, Marconi sails, and carefully streamlined hull are formed in this century's America's Cup racer (Figure 18). The best speed made by a very able representative of this class, the *Yankee*, was about 13.5 knots. It may not sound impressive compared with the *James Baines*, but no square rigger of the *Yankee*'s size could have come near it in the same wind. However, such a boat has no purpose other than to race with boats of its own kind. As far as speed is concerned, the America's Cup racer merely represents the attainment of another

FIGURE 17

The Paraglider

Close examination of the "flexwing" supporting this aircraft will show that each wing half is the exact counterpart of a lateen sail, even to the inclusion of the mast. An additional irony related to this turn of affairs is the present attempt to apply airplane wings to sailboats. The concept of the paraglider is attributed to Francis M. Rogallo of the National Aeronautics and Space Administration. The particular design shown in the illustration is the work of the Ryan Aeronautical Company.

FIGURE 18

(Photograph by Rosenfeld)

America's Cup Racers

These highly-refined racing machines (J-boat class) are the last word in single-hulled displacement sailboats. In the course of their development much of the sea-keeping ability of early contenders for the prized cup has been sacrificed. When rigged for racing they are not suitable for crossing an ocean, as did the *America*. The *Yankee*, shown here, holds the record for this class—13.5 knots. J-boats are much larger than the 12-meter boats, *Weatherly* (U.S.A.) and *Gretel* (Australia), which competed for the America's Cup in 1962 at Newport, Rhode Island.

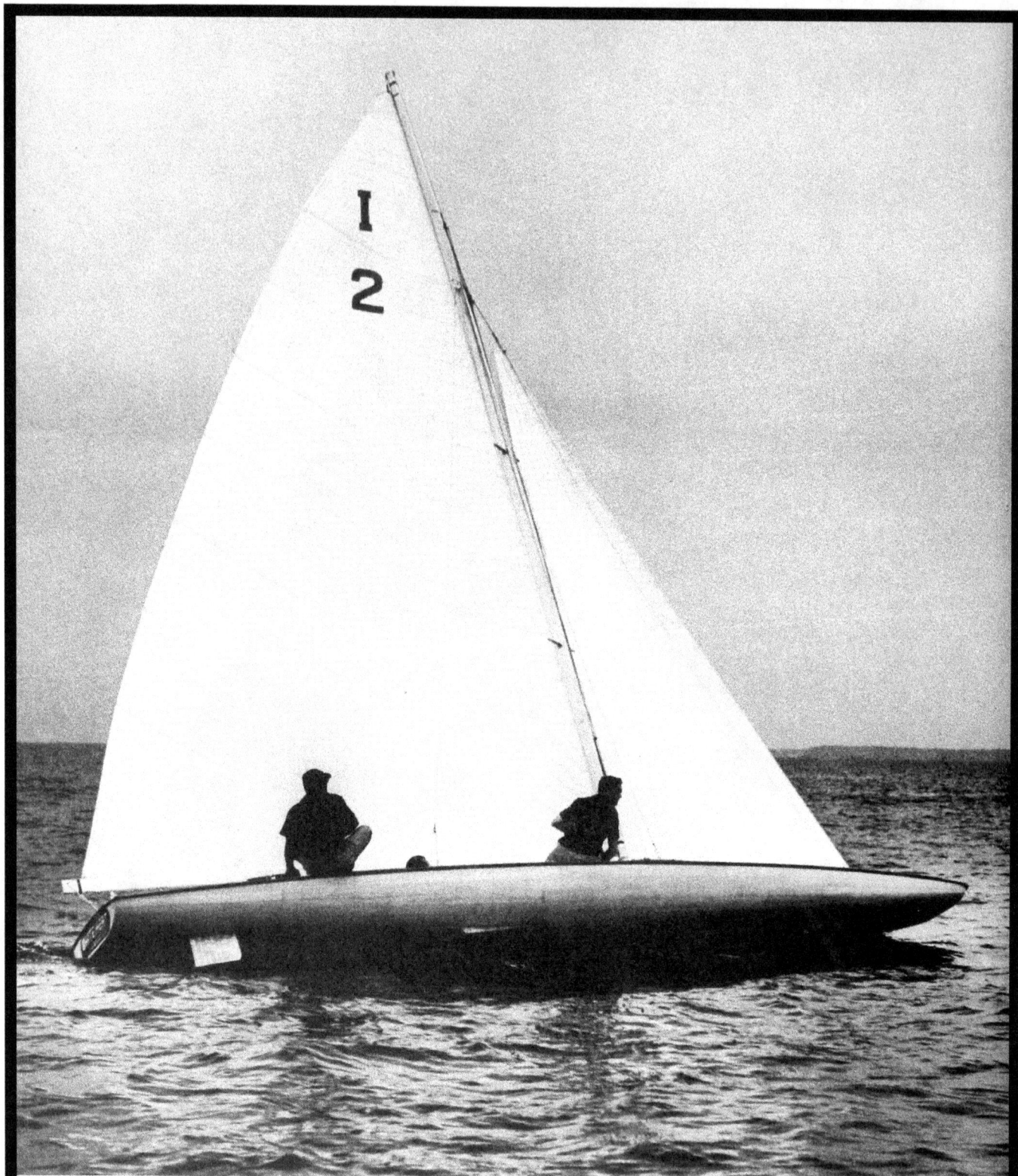

FIGURE 19

Sailing Scow

The scow is really two hulls joined together. This design provides ample deck and shallow draft, but rough riding in a seaway. This boat is deliberately sailed on its side so as to reduce the wetted area. It is faster than other single-hulled sailboats on all points of sailing, and it can plane when sailed directly downwind in a high wind. The scow in the picture has its windward bilge board retracted.

kind of limit, for certain sailboats, beyond which it is impossible to make substantial progress.

Every imaginable variation of hull shape has been tried in order to improve the speed qualities of single-hulled boats, but with little beneficial return. Two that have made some gains, the scow and the planing sailboat, are not comfortable ocean boats. The scow is a disguised catamaran, with the space between the hulls planked in. As shown in Figure 19, it sails on one of its bilges to reduce the hull's wetted area. A claim of 25 knots has been made for the scow. While it is hard to credit this boat with such speed potential, it is not difficult to assign to it a capability, on some points of sailing, equal to the catamaran, for which 25 knots is also claimed. According to the story I have received, the scow that made this record was sailing in a quartering wind and was planing, a feat few catamarans can match.

Fully planing sailboats are reputed to have better sea-keeping characteristics than scows, but in this respect they are still not equal to the best displacement boats, which are, after all, only a little slower. As a matter of fact, until the wind is strong enough to bring them to planing speeds, they can be beaten by a well-designed displacement boat. (However, the scow, which is the fastest single-hulled displacement boat, can plane on occasion and walk away from the fully-planing sailboat.) The planing boat tries to reduce its wetted area by riding on its stern, which is flatter than that of other sailboats (Figure 20). This kind of behavior does nothing to make sailing happier in rough water, but it does produce lift, which reduces displacement. Unfortunately, accurate speed records for the planing sailboat are not available.

Revival of Interest in Multiple Hulls

It is difficult to imagine how the single-hulled boat could be further modified to im-prove its speed. All avenues seem to have been explored and exhausted. Consequently, boat designers who have sought faster sailboats have been looking with more interest at the primitive but speedy double-hulled craft of the lonely Oceanic islands. These boats were developed under a unique set of conditions in that they are not descended from inland water boats. It is fairly certain that they were indigenous to ocean beaches right from the start. Therefore they had to meet the problems of beaching and open-water sailing simultaneously, which they did exceedingly well.

When first seen by white men over 400 years ago, the sailboats of the outlying Pacific islands were all slender and multiple-hulled, either catamarans or outrigger canoes. None depended on a low center of gravity for stability. The most famous was the "flying proa," developed in the Marianas Islands (Figure 21). This remarkable boat was reputed to go faster than the wind. It may very well have done so, in which case it was truly faster than any ocean-going sailing vessel developed before the 1950's, including the clipper ships and the America's Cup racers, which made their records in high winds—not by exceeding the speed of the wind.

The Remarkable Flying Proa

The first accurate description of the flying proa was furnished by the English admiral Baron Anson in 1742, although Magellan's chronicler reported them in 1521. Anson, who was a careful observer, judged their speed to equal or better the wind's speed and estimated their top speed to be 20 miles an hour, or about 17.5 knots. Since then, speeds of 18 knots have been verified beyond doubt.

The flying proa is worthy of some detailed attention at this point because in many respects it incorporated a fair number of radical features that, strangely enough, were forerunners of improvements arrived at recently by the use of modern scientific methods.

The Marianas Islands extend along a line that lies roughly perpendicular to the prevailing winds of the region. With rare insight, some early genius recognized that running with the wind and beating to windward were uncommon points of sailing in going from one island to another. Hence, a boat appropriate to the traffic would be one designed primarily for reaching (sailing with the wind more nearly from the side, rather than from ahead or behind).

Basically the flying proa consisted of a slender main hull deeper than it was wide, a shorter outrigger log braced to windward, and a triangular lateen sail raised from a central mast. The proa was double-ended and could sail equally well in either direction. The natives would sail from one island to the next, and, when ready to return, they would turn their own bodies about to make what had been the stern, the bow. The maneuver was completed by reversing the sail, through clever rigging, and sending the helmsman aft with his oar, although sometimes a helmsman was kept at each end to make the maneuver faster. Since the wind was always from the same side, ballast could remain on the outrigger when the boat's direction was reversed—altogether ingenious.

The details are even more interesting. The main hull showed a flat face to leeward and a rounded one to the outrigged side, so that a horizontal section of the hull had the approximate shape of a hydrofoil profile. The resourceful islanders had contrived a boat with no keel, centerboard, or leeboards, so that it was convenient for beaching, and yet it displayed good resistance to leeway.

Proa vs. Catamaran

The proa spread to other parts of the Pacific but never became popular with Western man. For corresponding size its sea-keeping qualities were as good as any built into European boats. The principle of rigging ballast on a long, horizontal arm, rather than in a short, vertical one, permitted a considerably lighter boat for equal payload. The flying proa did not suffer from

the clumsiness or the high torsional stresses of the catamaran, nor from the tendency of the double outrigger, or trimaran, to submerge its leeward float. It had, moreover, two other speed advantages over the catamaran. For equal displacement the proa had a longer waterline length; and, even more important, the unequal length of its hulls imposed maximum wave drag on each hull at different speeds. Thus, the shorter outrigger passed its maximum wave drag before the main hull reached its own maximum. A catamaran, on the other hand, is caught in the middle, with a doubled maximum wave drag at a speed between the proa's critical ones for the main hull and outrigger. Yet the catamaran and the trimaran, both inferior to the proa, are strangely enough the prime objects of present Western interest. (One California experimenter, however, is giving his attention to the proa and attempting to overcome the inconveniences of coming about.)

Coming About in The Proa

The need to reverse direction in order to come about, a matter of no inconvenience to the Marianas Islanders, may explain the unpopularity of the proa. This maneuver might seem hazardous, particularly in narrow waters, to a sailor trained to come about by keeping headway. (In a pinch a proa can come about on an opposite tack in the fashion of symmetrical boats. But it would develop more leeway with the wind on the main hull side, and the mast would create an inefficient discontinuity in the sail. Moreover, a leeward outrigger runs the risk of submerging in strong winds.)

The experienced sailor dislikes losing headway when changing from one tack to another because his boat does not respond to the tiller and is therefore out of control. He cannot think of a sailboat as being other than "in irons" when it comes to rest while changing tacks, because he associates the operation with facing into the wind. But the proa never runs the risk of being in irons when it comes about.

FIGURE 20

(*Photograph by Rosenfeld*)

Planing Sailboat

Planing action was the first lift principle applied to sailboats. When the speed is high enough, an upward force is developed by the flat planing surface, which serves to reduce the draft and therefore the drag. Until planing speeds are reached, which usually requires strong winds, the planing sailboat is no faster than the well-designed displacement sailboat. The boat in the picture is moving with the wind on the quarter.

FIGURE 21

Marianas Island's Flying Proa

These fine canoes no longer sail the Pacific. An inferior modern version is now found in the islands of Micronesia. The main hull of many early proas was 40 feet long. Its flat leeward face gave it excellent windward performance, and its slender lines allowed the highest speed of any small, ocean-going sailboat until well into the twentieth century.

It can reverse its direction very quickly by reversing its sail, and it never loses driving power in the process. Moreover, when sailing downwind, the proa need never be subject to an accidental jibe, a situation feared by sailors of ordinary boats more than being in irons. In going from one broad reach to the other, the proa merely reverses its sail, keeping the luff always to windward, and then changes to the new downwind direction without ever being dead before the wind. In this way it can sail downwind faster by tacking on broad reaches than any other boat can by going full with the wind.

On the other hand, the objection to the proa may be more fundamental in nature. It may simply stem from an aversion to asymmetry. Western man has worshipped symmetry for so long, he must find it abhorrent indeed to accept the thought of a machine without bilateral symmetry. I must confess that in my work with asymmetrical sailboats I could not rest, even when their performance surpassed all prior models, until I had designed a version that had at least a token of bilateral symmetry. In truth the symmetrical boat was aesthetically more satisfying, but the symmetry existed only in the transition from one tack to another.

The flying proa, as good as it was, is now almost forgotten. Whatever the reason, surviving proas in the South Seas are inferior to it. The present renewed Western interest in the double-hulled boat is mostly confined to the catamaran. With every modern refinement in sails, hulls, centerboards, and construction materials incorporated into their designs, these newer catamarans may occasionally exceed the speed of the wind on a broad reach. Speeds of 18 knots have been recorded for boats similar to the one shown in Figure 22.

One claim of 25 knots has been made for the catamaran, but I would guess that some special set of conditions, such as a 30-knot wind coupled with fairly calm seas, was largely responsible. I cannot see how a boat of this kind could move at this speed in even a moderately rough sea without tripping itself on a wave and submarining, or else breaking up under the high torsional stresses to which catamarans are notably subject in uneven waters. (However, the advantages of modern materials and engineering are overcoming this problem.) Be that as it may, such speeds are perhaps the extreme performance limits possible with the catamaran, and then only under very special conditions; *i.e.*, high wind, calm sea, broad reach, all the movable ballast hanging out to windward—and great danger of upsetting the boat at any moment.

The "Sailplane"— a Modern Departure

Thirty years ago Malcolm McIntyre developed a double-hulled boat that made similar speeds without going through the same torture. It was cleverly designed to generate a righting moment equal to the overturning moment produced by the wind, and it employed a system of flexible joints between hulls that removed the torsional stresses associated with the catamaran. The boat, which was called a "sailplane," is shown in Figure 23. The sailplane set a record of 20 knots. Unfortunately it was born in the Great Depression, at a time when catamarans were not in favor, and, more unfortunately, it was considered a catamaran mutant, which it was not. Moreover, it arrived on the scene after famed yacht designers like Nat Herreshoff had tired of challenging the racing committees that banned the catamaran. Herreshoff's own catamarans had many novel features, including a system of flexible joints that anticipated the sailplane's system. But as clever as Herreshoff was, he evidently did not possess the great gift of creativity that McIntyre brought to bear on the problem of the sailboat.

A Meeting of Minds

McIntyre, who was a man after my own heart, anticipated the ideas of many who followed him, including myself. Until I entered into correspondence with his daughter, Mrs.

FIGURE 22

Catamaran

During the last ten years the catamaran has risen sharply in popularity. Many marine architects believe that it combines speed and seaworthiness better than any other boat. On broad reaches in strong winds, boats like the one shown above have attained speeds of 18 knots.

FIGURE 23

Sailplane

The sailplane, introduced during the 1930's, was not a catamaran, and, unlike the catamaran, it did not depend on ballast to remain erect. The hulls were connected by jointed outriggers, which eliminated torsional stresses. The inclination of the two centerboards and the two sails was such as to develop an upward force on the lee side and a downward force on the weather side when the boat sailed on a tack. Thus, restoring moments were created that were proportional to the strength of the wind. The sailplane had higher speed potential than the catamaran.

Louis Darling, I thought that the kite principle, which was applied to the last aerohydrofoil I designed, was uniquely my own. Forty years ago McIntyre was thinking fundamentally in the same way. In reading his published articles I was struck again and again by the amazing coincidences between his work and mine—the same questions, the same approaches. "A great deal of time, brains, and money have been spent in trying to perfect . . . the ballasted type of racing boat; but are they so much better now than they were in the beginning? . . . These boats have been developed about to the limit. . . . Something new in principle will have to be gotten up to get really high speed under sail. . . . The ideal theoretical fast sailboat should be one that will stand upright in any breeze, have no ballast in any form and only enough displacement to comfortably float the crew, sails, and rig." In describing an early model he invented, his words were: "This was a light skimming dish hull, with a vertical centerboard and a sail held out to leeward at an angle of 45 degrees. This sail lifted like a kite and was so placed as to cause no overturning moment."

Fate cheated Malcolm McIntyre in many ways. If he had been born a generation later he would not have failed to incorporate hydrofoil lift principles as well as airfoil lift principles into the ultimate unbeatable sailboat. As a matter of fact, he was using the centerboards in his sailplane as hydrofoils, but, unfortunately, for negative lift as much as for positive lift. Nonetheless he achieved a record that stood for a quarter century. I suppose some day someone who never heard of the sailplane or of McIntyre will reinvent it and demonstrate once again that it is faster and better than the catamaran.

The Monitor

Impressive as the sailplane is, higher speeds have been made with another kind of boat, one that has taken a long stride toward becoming neither a single-hulled nor a multiple-hulled boat, but more nearly a zero-hulled boat. Figure 24 shows this boat, the *Monitor*, in action. Running on a broad reach with the help of its ladderwork of hydrofoils, the *Monitor*, invented by F. G. Baker, an experimenter with hydrofoils, was able to lift its hull above the water and attain 30 knots. It made this speed without being tossed all over the water by the least wave and without any great hiking of ballast after the hull left the water. But like any machine that is the first of its kind, it suffered from a number of defects.

It was, first of all, extremely complex; indeed so complex that its cost would rule out a general acceptance by the average sailing enthusiast. Furthermore, it had no real advantage over displacement boats in sailing upwind, or on any other point of sailing in light winds. The foils could not lift the hull clear of the water until the wind was at least 13 knots, and, until this speed was reached, little righting moment was developed by the foils. At about this speed the boat was neither fish nor fowl, necessitating the same old acrobatics on the part of the crew to keep it erect. Nevertheless, without the foils this small boat's speed would have been limited to less than a third the speed it actually achieved.

The great difficulty in comparing the speeds of clipper ships, America's Cup racers, scows, planing sailboats, proas, catamarans, and hydrofoil sailboats is that the speed record is almost never accompanied by information on wind speeds or what the boats would have done on other points of sailing. There is also the added difficulty of trying to compare boats of different lengths and weights, which nearly always involves debatable correction factors. In general, the performance is reported only when a record is broken sailing downwind in a howling gale. As any sailor knows, the critical problems of the sailboat are more closely associated with the conditions of upwind sailing in light breezes.

FIGURE 24

Hydrofoil Sailboat

The *Monitor* represented the first concerted effort to lift the sailboat's hull completely out of the water. The complicated ladderwork of foils performed this feat when the wind exceeded 13 knots. To keep the boat from pitching, it was necessary to couple the rear foils to the mast. As the force on the mast increased, mechanical linkages served to reduce the angle of attack. The *Monitor* holds the speed record for sailboats—about 30 knots. The above photograph, showing her in high-speed motion, was supplied by the inventor, F. G. Baker.

The 40-Knot Sailboat

Limitations of Classical Design

By one scheme or another, and after centuries of work, men have coaxed a few extra knots out of the reluctant waters. The last few knots seem to have taken all that modern technology has to offer. Yet the sailboat still does not move upwind at motorboat speeds. Even when iron or lead ballast is removed, and multiple hulls are employed instead for stability, sailboats still do not go very much faster upwind. Nor is much more speed obtained when we use a planing hull, the form that makes the fast motorboat possible. Nor have we seemed able to transfer the application of hydrofoils from motorboats to sailboats without paying a great price in complexity for the occasional high speed. Moreover, somewhat like that of the planing hull, the hydrofoil's contribution to high speed is not really effective in increasing the lower speeds sustained in light winds. All these boats, as well as the earlier types, would merely capsize if we attempted to add the amount of sail required to drive them to windward as fast as the wind.

Over four hundred years have passed since civilized men witnessed primitive boats making 18 knots. Since then, with our vastly increased knowledge, and a fabulous new assortment of materials with which to work, only about 10 knots have been added to downwind sailing by a few highly specialized boats. None has attained improved upwind speeds to the same extent.

I suspect that the historical attempts to improve the sailboat have been akin to the process of trying to improve a horse in order to get an automobile. Without a fundamental reappraisal of the barriers to higher speed, the best result to be expected from this incremental improvement procedure is a sea-going analogue of the camel, which is an improvement over the sea-horse, but a far cry from a sea-automobile.

Three years of my spare time have been furtively spent in the reappraisal. Every discipline entering into the design of a very fast sailboat had to be re-examined for built-in misconceptions, a number of which were uncovered; and with these cleared away, new design concepts have at last appeared to give hope for a quantum jump in sailboat speeds, on all points of sailing.

PART TWO

History of the Aerohydrofoil

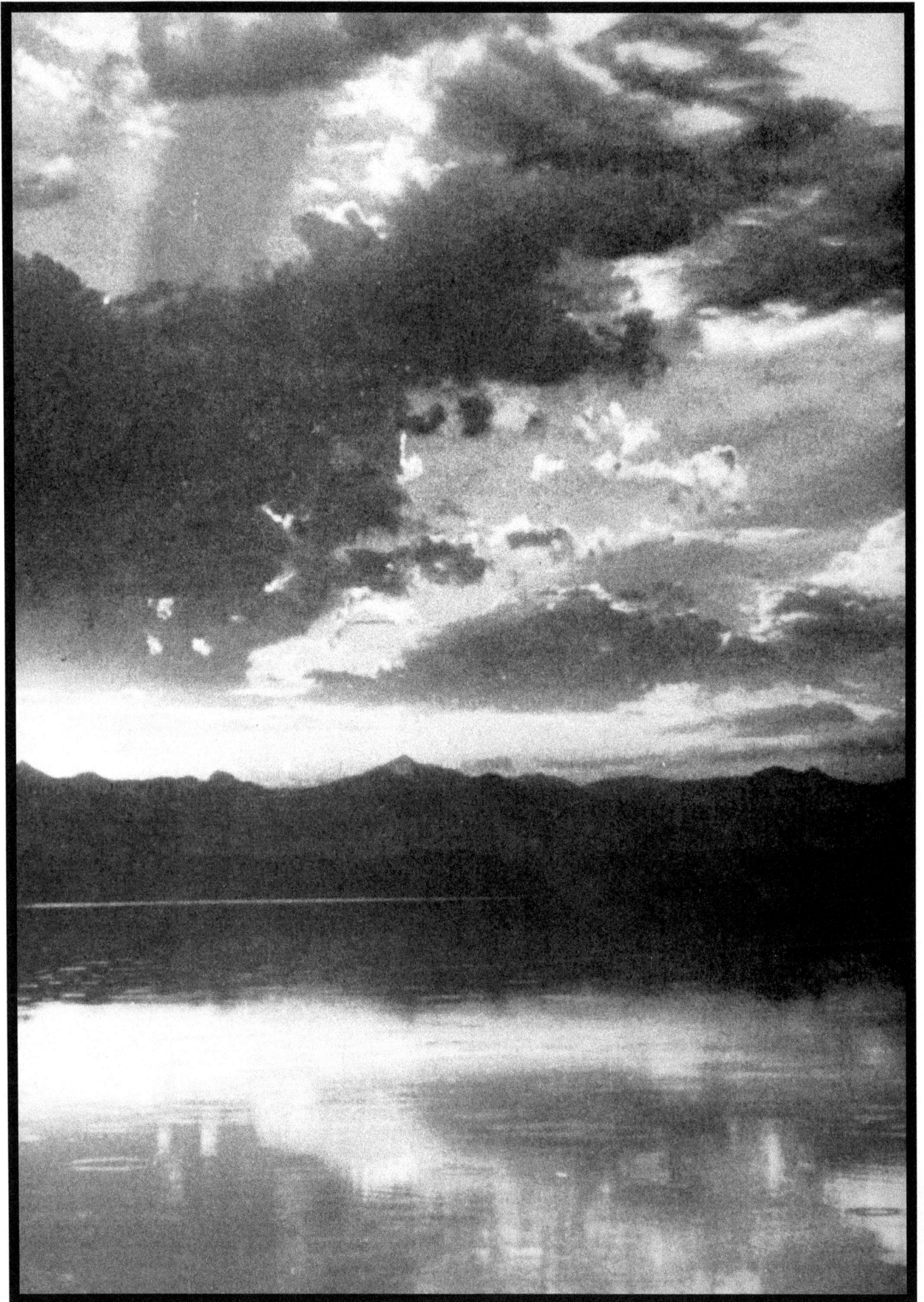

CHAPTER IV

The China Lake Inventions

If there is any order to these pages, little of it was evident when I started work on the aero-hydrofoil. I knew of no one working on a similar device (and there were some); I made no literature surveys; I had no well-developed theory. All of this came much later, after I had experimentally tested some of my hunches. When I found that I had had the good fortune to guess correctly a few times, and saw clearly the first real indications of something unique, I felt the time was proper for evaluating the worth of the idea. Only then did I undertake the analyses discussed in Part III of this book. My purpose in describing the unorganized nature of my first attack is to be sure that the reader is not misled into thinking that a planned, formalized approach is the direct path to a new idea. On the contrary, formal methods of synthesis and analysis may be excellent for attacking standardized problems, but they seldom, of themselves, lead to new concepts.

My interest in the hydrofoil sailboat was stimulated by, of all things, living in the Mojave Desert, where for twelve years I worked on Naval rockets and missiles. Being land-locked and dry at the Naval Ordnance Test Station (NOTS), China Lake, California, apparently provoked me into thinking about water. (I understand that the Navy's most profitable recruiting efforts occur away from coastal regions.) There was a dry lake within sight of my house, and every time it filled with winter rains I gazed at it long and hard, wondering what could be done with this ethereal little body of water before it vanished—often in a matter of hours (Figure 25). Obviously it was not big enough or deep enough for a man-carrying boat, but what an ideal model basin it was. The water was so uniformly shallow, one could walk or run with a small model all over the lake.

Mirror Lake, as it was called, rarely reflected anything even when it contained water.

By the time storm water reached the dry basin it was muddy, laden with debris, and ruffled by the same wind that brought the storm. Its most remarkable tenants were the brine shrimp, known locally as "fairy shrimp." These little animals hatched from their tiny eggs in a matter of days after the arrival of the winter rains, and within a few weeks they were several inches long; whereupon they would immediately proceed to develop and lay another generation of eggs. The eggs would become completely dehydrated in the 120-degree heat of summer and were then spread by the wind from one dry lake to another all over the Southwest. Thus did my predecessors at Mirror Lake establish the tradition of using the wind for transportation.

In winter a few hardy hobbyists would venture out to test propeller-driven model boats on the lake. The air propellers and their gasoline motors were noisy, cantankerous little monsters with about the same nuisance value as airplanes. I could never seem to get interested in such intruders on the desert peace. Stirred by my own resentment, I wondered if a model sailboat could do as well, or perhaps better, in the brisk and steady winds that blew across the lake in the wintertime.

Path to New Ideas

Obviously a keeled boat, unless it were microscopic, would have too much draft to clear the bottom. Except in a few small areas, rarely more than an inch or two of water covered the lake. Whatever device was to resist leeway had to be near the surface—more like a leeboard than a centerboard. Also, since a deep, weighted keel was forbidden, static stability would have to come from widely spaced floats. It occurred to me that perhaps the function of the centerboard and the floats could be combined in the form of a buoyant hydrofoil. I reasoned that if the hydrofoils were large enough and inclined at suitable angles, one could get static stability, the proper resistance to leeway, and lift to raise the boat out

of the water and reduce the drag. Not the least of the attractions was the novelty of eliminating the hull.

These considerations led to a number of designs similar to the two-foot model shown in Figure 26. The first one I tried was a terrible disappointment. It promptly capsized as soon as it was struck by the first gust. In the short time it took the model to flip over, my mind flashed back to Archimedes' principle of buoyancy, laboriously learned in freshman physics. I had ignored the fact that the leeward foils would have to have a potential buoyancy equal to the total displacement of the boat; otherwise, as soon as the windward foils lifted out of the water, the leeward foils would submerge, leaving no restoring moment. Dynamic forces could produce the necessary moments once the boat was in motion, but at rest the foils needed more freeboard. When this was supplied, there was no more trouble from that source.

All in all I constructed eight symmetrical models, some after I had obtained much better results with asymmetrical versions. Although I knew that it was impossible for a sailboat to get superior performance with symmetrical hydrofoils, I never really understood why until I conducted a more careful analysis with more elaborate models at Newport, Rhode Island, over a year later. Only then did I give up the quest for absolute symmetry.

The symmetrical models were very fast on a reach, but on a beat they were almost worthless. The leeway was considerable then, and the speed low. On this point of sailing they were no better than ordinary sailboats. The explanation is detailed in Chapter X.

Inclination of Hydrofoils

The usual "V" inclination of the hydrofoils applied to motorboats is fine when side forces are not at work. All the foils lift as they should and the side components are mutually cancelled. This arrangement also works well for sailboats going with the wind or on a broad

FIGURE 26

Symmetrical Aerohydrofoil

This model was typical of the first hull-less sailboats tested. In plan they resembled iceboats. Not furnished with an airfoil to replace the sail, or with hydrofoils inclined in the same direction, they did not beat to windward any better than ordinary boats. They did, however, develop unusual speed on a broad reach.

FIGURE 27

Asymmetrical Aerohydrofoil

Inclining all foils to provide, simultaneously, both lift and resistance to leeway gave this model much better performance on all points of sailing. The two-foot craft sailed a broad reach at 14 knots in a 10-knot wind. It was also able to beat to windward as fast as the wind when pointed off the wind about 70 degrees. Its ability to sail upwind was limited by the large angle of attack required with an ordinary sail.

reach. Unless there is a centerboard (or its equivalent in the form of struts), however, the foils must supply the leeway resistance essential to beating. Very little resistance is developed until either the angle of attack or the immersion of the leeward foils becomes much greater than that of the windward foils. This means that the windward foils cannot effectively supply either lift or resistance to leeway —for the most part they can only add drag.

The solution is to incline all the hydrofoils in the same direction. The model in Figure 27, two feet long, was one of a number designed to include this feature. In some ways it was like the proa: it changed tack by reversing direction, and its ballast, carried on the windward spar, did not have to be shifted for each tack. The three foils coincided with the extremities of the proa, but in all other respects the resemblance to the proa's hulls was slight.

Taste of Success

The thrill I experienced when testing this model was equal to the exhilaration of my first rocket firing. When at last I had the foils and sail properly adjusted to give some directional stability, it climbed on its feet and scampered on a tack much faster than I could chase it. I caught the little runaway only after it had turned directly into the wind and slowed down. I called for my wife to come and witness what I had done, and when she arrived I set the boat in motion again. This time I pointed it a little more off the wind. It flew so quickly it barely touched the water. I could only watch in astonishment as it sped across the lake, continuing its motion for 15 feet up the opposite bank. The foils were destroyed but I didn't mind. This looked like the answer to my dream—a boat that could trim the propeller-driven models!

To get some idea of the relative speed of the boat with respect to the wind, I used, at first, an unorthodox but simple method. Later I checked the speed with more conventional systems and found my first calculation quite ac-

curate. I would run with the boat at the boat's speed, then turn and run at the same speed directly downwind. If I felt a breeze on my face, the boat was faster than the wind; if I felt no breeze, the boat was moving at the wind's speed; and if the breeze came from behind me, the boat was slower than the wind. Under certain conditions I was able to get some numbers into these estimates. My top speed (then), checked against an automobile, was 14 miles per hour. When the boat moved at my top speed or greater, and the wind at my top speed or less, I could be quite positive about actual speeds. During the one test in which I used a wind indicator and the boat was timed between markers, the speed of the wind was 10 knots while the boat, at about 70 degrees off the wind, was sailing at 12 knots. Visual estimates based on this kind of training convinced me that the little boat often reached and even exceeded 14 knots in 10-knot winds.

My best friends and most trusted confidants during these days were high school students who haunted the lake whenever they were inclined to avoid their studies. The most enthusiastic was Peter Vicens, an American Field Service exchange student from Spain. He literally jumped for joy whenever the little boat skipped across the waves. Peter, a sailor himself, is now back in Spain awaiting the day when he, too, can work on a man-carrying aerohydrofoil. I hope this book will speed that day.

Burden of The Scientific Approach

These experiments were watched with interest, and with some amusement, I suspect, by residents near the lake. Dr. Ivar Highberg, head of the Test Department at NOTS, was one of the observers. He promptly classified the activity as fundamental research; and, as president of the local chapter of the Research Society of America, he made sure I became a member in good standing without delay. This heavy burden of research responsibility was almost my undoing. It made me feel that I should be more scientific in my approach, so I consulted a hydrodynamicist. I was immediately told that the scheme violated well-established facts about hydrofoils, and that even if thick hydrofoils worked on the models, which he found surprising, they would certainly not work on full-sized versions. According to this expert, such foils would lose lift at high speeds; and so they would. What he failed to consider is that the thick parts of the foils are out of the water at high speeds, and the thinner, immersed portions do not lose lift. My consultant could attack the total problem only as a collection of partial problems, not as an integrated one. Therefore he was unable to help me realize the maximum potentials of the various parts.

My experience with marine architects was not much happier. I have now learned to tell when they are about to display a complete knowledge of all the many things that *cannot* be done. It is when I am indulgently asked, "What happens when your high-speed sailboat hits a log?" or "How will it behave in a state five sea?" They, of course, know as well as I do what happens, or they would be poor marine architects. If anyone persists in striking a sufficiently large log or wave at full tilt with any boat, now or in the future, the results can be catastrophic (or, at the very least, ruin one's entire day, as my friend Colonel Jack Wade has put it).

There is no sure way to avoid impacts with ponderous solid or liquid obstacles, except to have wits enough to avoid them or prudence enough to slow down when approaching them. Wits and prudence belong to the pilot, not to the boat, although some boats can handle a pilot's mistakes better than others. One never knows which is which until a boat is tested under critical conditions—a situation few marine architects are able to duplicate on a drafting board.

A Sympathetic Ear

Fortunately, one hydrodynamicist, Tom

Lang of NOTS' Pasadena Annex, was evidently more careful in his review, and through his help I was able to conduct tests with a full-scale foil in a water tunnel. The results convinced us both that the foils would scale up and would be fairly free of ventilation at high angles of attack, at least up to the tested speed of 17.5 knots.

Many varieties of foils were tried on the asymmetrical models without achieving any pronounced difference in their performance. But I did learn a great deal about how to suppress bow waves and about the sweep angles needed to shed seaweed from the leading edges of the foils. This did not really compare in importance to what I learned about one model from motion pictures that were obtained through the kind assistance of Richard Carlisle, a colleague at NOTS. The pictures showed quite clearly that there was greater immersion for the front foil than for the rear one. A rough estimate of its depth indicated that occasionally it must have been dragging in the mud bottom of the lake. This did not bother me nearly so much as did my negligence in not preventing it.

I had made the mistake of applying to this boat the same reasoning one would apply to a hull. Owing to the bow wave produced by an ordinary hull, an upward pitching moment is developed to counter the downward pitching moment exerted by the sail when it drives the boat. There was no correspondence to the hull's natural restoring moment in the hydrofoil boat. The correction could lie either in increasing the angle of attack of the front foil or in redesigning the boat so that the resultant moment from the sail would lie perpendicular to a line connecting the leeward foils. The second approach, which at first looked more complicated, turned out to be simpler and better in the end. It required holding the sail almost parallel to the line connecting the leeward foils, and *turning the foils* to get the desired sailing direction.

At this point I undertook to apply for a

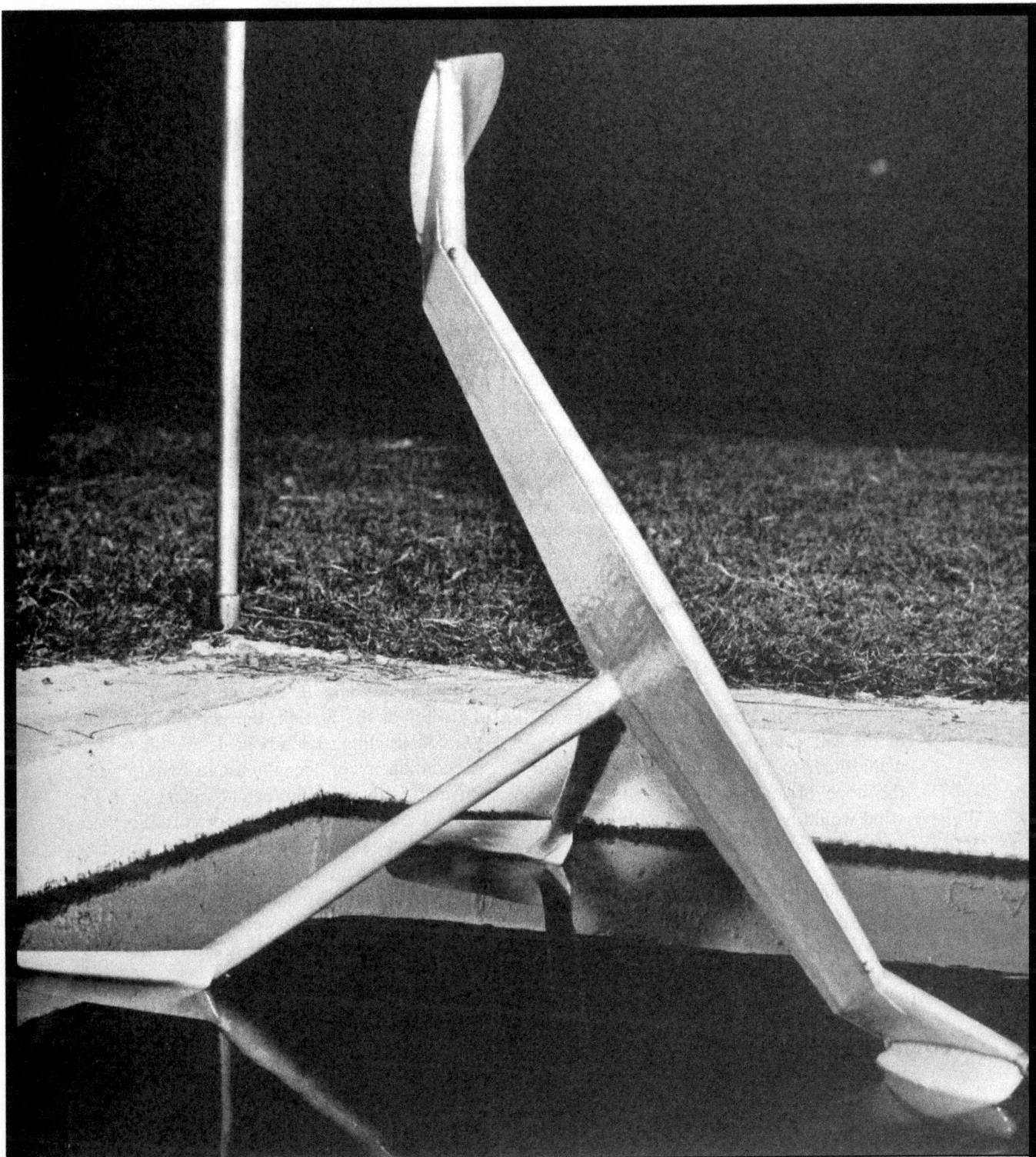

FIGURE 28

True Aerohydrofoil

The true aerohydrofoil comprises only airfoils, hydrofoils, and a connecting structure. Although the theoretical potential of such a craft is high, practical design problems interfere with the full use of the foils. The model shown represents an early attempt to solve the problem of coming about by tilting the entire frame. It was able to beat upwind at 30 degrees off the wind, but it was too heavy and slow. Note the use of the air rudder.

patent on my version of the hydrofoil sailboat, anticipating all sorts of overlapping with other applications. The first search at the U. S. Patent Office, however, uncovered no duplications; and even more astonishing, was the discovery that there had been no claim at all for a hull-less boat. The patent files were bulging with inventions that combined displacement and dynamic lift for airborne bodies (*e.g.,* dynamic-lift blimps, gas-filled wings), but no one had applied such combinations to a device that worked in water, where, it seemed to me, it would make much more sense. The buoyant lift of a foil in water is eight-hundred times its buoyant lift in air. Moreover, this lift is realized in water even though the foil is filled with air, whereas under the same conditions no static lift is realized in the atmosphere. Here we have an example of another scientific lag, but in this one the nautical experts trailed the aeronautical experts.

Application of Airfoil

The problems now converged on the design of the sail. I could see that sooner or later I would have to attempt the use of an airfoil, or the hope of making high-speed progress upwind would be vain. In line with this thinking, I constructed four true aerohydrofoils, one of which is shown in Figure 28. This one did indeed beat to windward at 30 degrees off the wind, but not at the desired speeds. It was too heavy and would not lift itself high enough out of the water. Therefore, in addition to more than enough resistance to leeway, it had more than enough drag, but it represented my first attempt to get directional stability with an air rudder. These models depended solely on tilting to go from one tack to the other. This, of course, contributed nothing to practicality.

Again I went to the specialists for help, this time to aerodynamicists. Again I found myself drawn into arguments over what could not be done. The issue concerned my contention that airfoil analysis as applied to an airplane wing

is not applicable to a sail, even if the sail should be an airfoil. Lift and drag, I claimed, could not be applied to a sailboat in the same way it was applied to an airplane. As a matter of fact, airfoil drag was a penalty only when sailing very close to the wind; otherwise it helped to drive the boat! I would win the argument on the basis of paper proof, which neither endeared me to, nor really convinced, the first few aerodynamicists who were subjected to the ordeal. Eventually I had to develop the theory into a coherent form by myself, and thereafter I had less trouble asking for specific information from the aerodynamic community.

Best Materials

None of the other true aerohydrofoils I tried had the all-around qualities of the proa-like versions. I decided, therefore, to defer looking for more cake and started construction of a man-carrying model on the proa plan. Long before making this decision, I had discussed with Dr. William B. McLean, the Technical Director of the Station, the possible materials for constructing such a boat. It was his recommendation to consider the use of foam plastics covered with resin-impregnated glass cloth for all the parts. In the course of forming these materials I came closest to giving up the whole venture in defeat. The process deteriorated into one of developing materials, rather than a boat. But the methods were learned and I am now convinced that there are no better substances for the aerohydrofoil.

In the midst of all this, my regular work assignment was changed to duty at the Naval War College, Newport, Rhode Island. It then became a mad race to finish the man-carrying boat and to test it before I left California. For this crash program I had the best help anyone could ask for in the person of Tony Hines, a pattern-maker at the Station. His skill and perseverance met every problem.

We did finish the boat in time to try it twice on a mountain lake under rather poor circum-

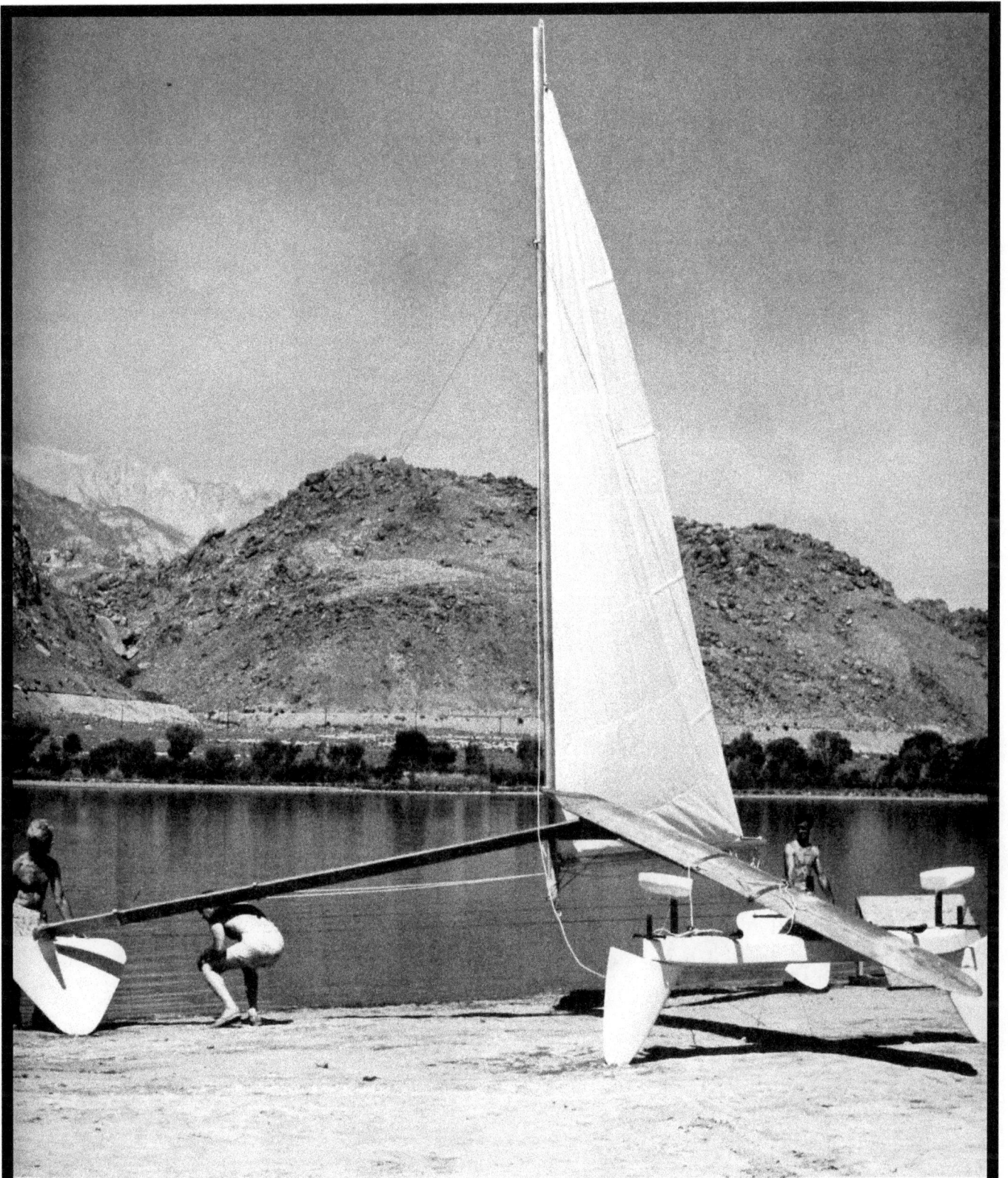

FIGURE 29

Man-Carrying Aerohydrofoil

(*Ready to Launch*)

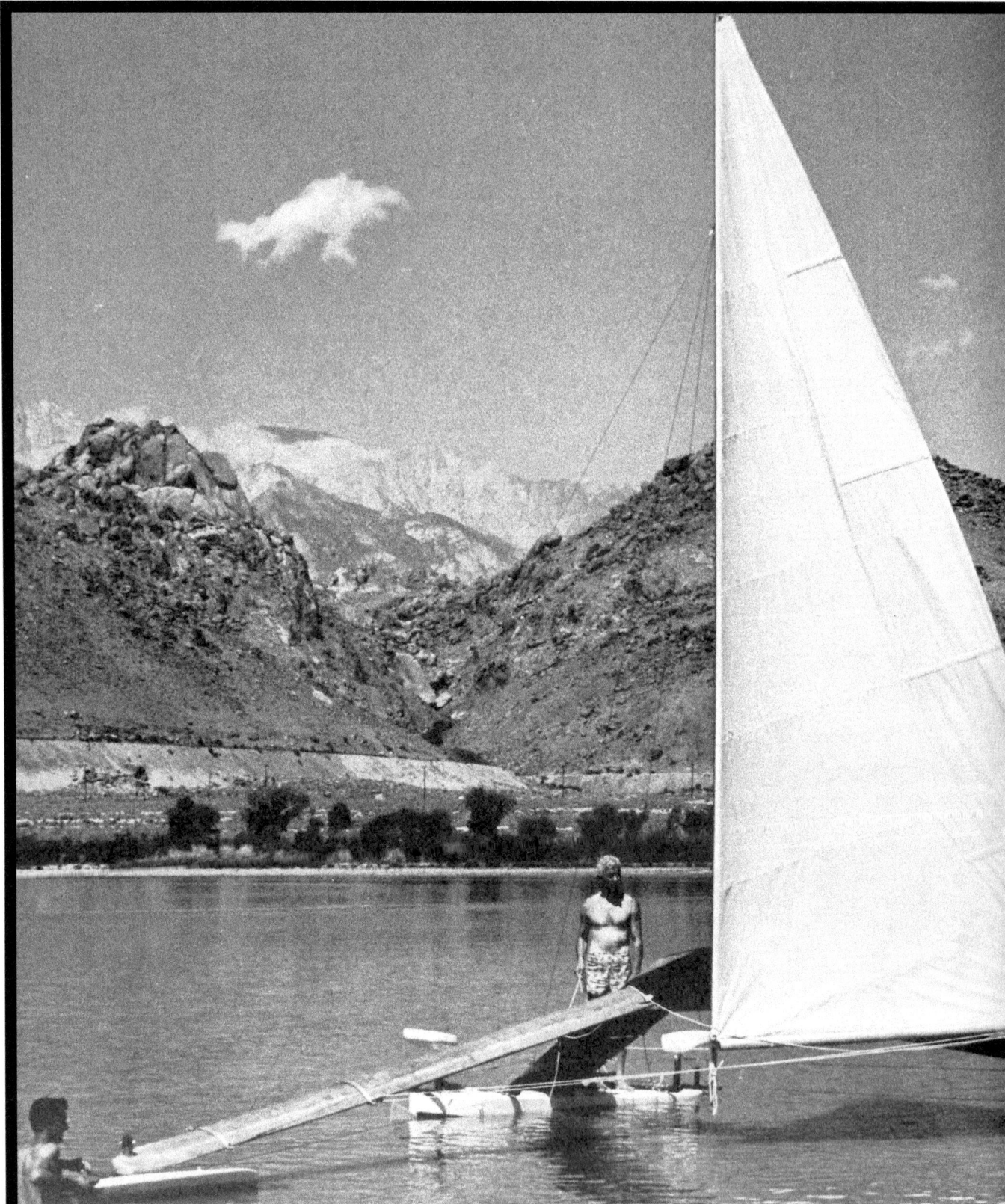

FIGURE 30

Man-Carrying Aerohydrofoil

(No Wind)

The author is standing on the windward float. Tony Hines is helping, with an assistant,
to hold the boat in readiness for the expected wind—which never came.

stances (Figure 29). I learned very little from these episodes, aside from the fact that I had made more atrocious mistakes on the large scale than ever before. My control system was inadequate and could not compensate for my misjudgment of the sail's center of pressure. In my ignorance I had assumed it would follow the one-quarter chord law of airfoils. Its true center of pressure was so far behind this value that I found it impossible to turn the boat off the wind enough to sail on a tack. This was one piece of information a good marine architect could have given me. But through my own self-defeating stubborness, I had given up consulting marine architects after the first few disappointing exchanges. No time was left to make the necessary changes. So there, at China Lake, still sit the components, and in Figure 30 you have a picture of a man at the brink.

CHAPTER V

The Newport Inventions

FIGURE 31
Symmetrical-Asymmetrical Aerohydrofoil (Port Tack)

In this view the leading edge of the airfoil is facing the observer. The wind is blowing directly into the picture, and the boat's motion is toward the lower left in line with the crosspiece.

Now that I recognize the deficiencies of the boat I almost sailed, there is, of course, no point in going back to the scene of the crime for a post-mortem. During the year I spent away from it, I had time to think more about the true aerohydrofoil; and, as if to give me a better perspective and deeper insight, fate let me gaze every day at the unending changes in the seascape about Narragansett Bay. In this inspiring setting I came at last to a concept that clothed the simple theory of the aerohydrofoil in practical raiment.

The year at the Institute of Naval Studies, attached to the Naval War College, was a year of free investigation wisely and generously given to me by the Navy, mainly through the efforts of Haskell G. Wilson, Associate Technical Director of NOTS. My mission at the Institute as a Naval Research Associate was, primarily, to make whatever contribution I could to a forecast of the Navy's long-range tactical problems and to refresh and update my knowledge of scientific and military affairs. In this endeavor I was joined by three other public servants, a social scientist, and six officers, each drawn from different parts of the Navy's far-flung establishment.

My tour at Newport accomplished much more than was scheduled, however. It restored my interest in the ocean. The rockets and the missiles and the satellites and outer space began to fade with each passing day. Soon I saw that the unknowns of the ocean were far more fascinating than those of space—and far more important to man's immediate needs. It was wonderful to discover again an almost untouched domain ripe for individual pioneering, and to leave for awhile the company of industrial spacemen who can undertake no project for less than a hundred million dollars and a thousand men, and who play with fire without knowing warmth.

Encouraged by Rear Admiral Edwin Hooper, Director of the Institute, I embarked

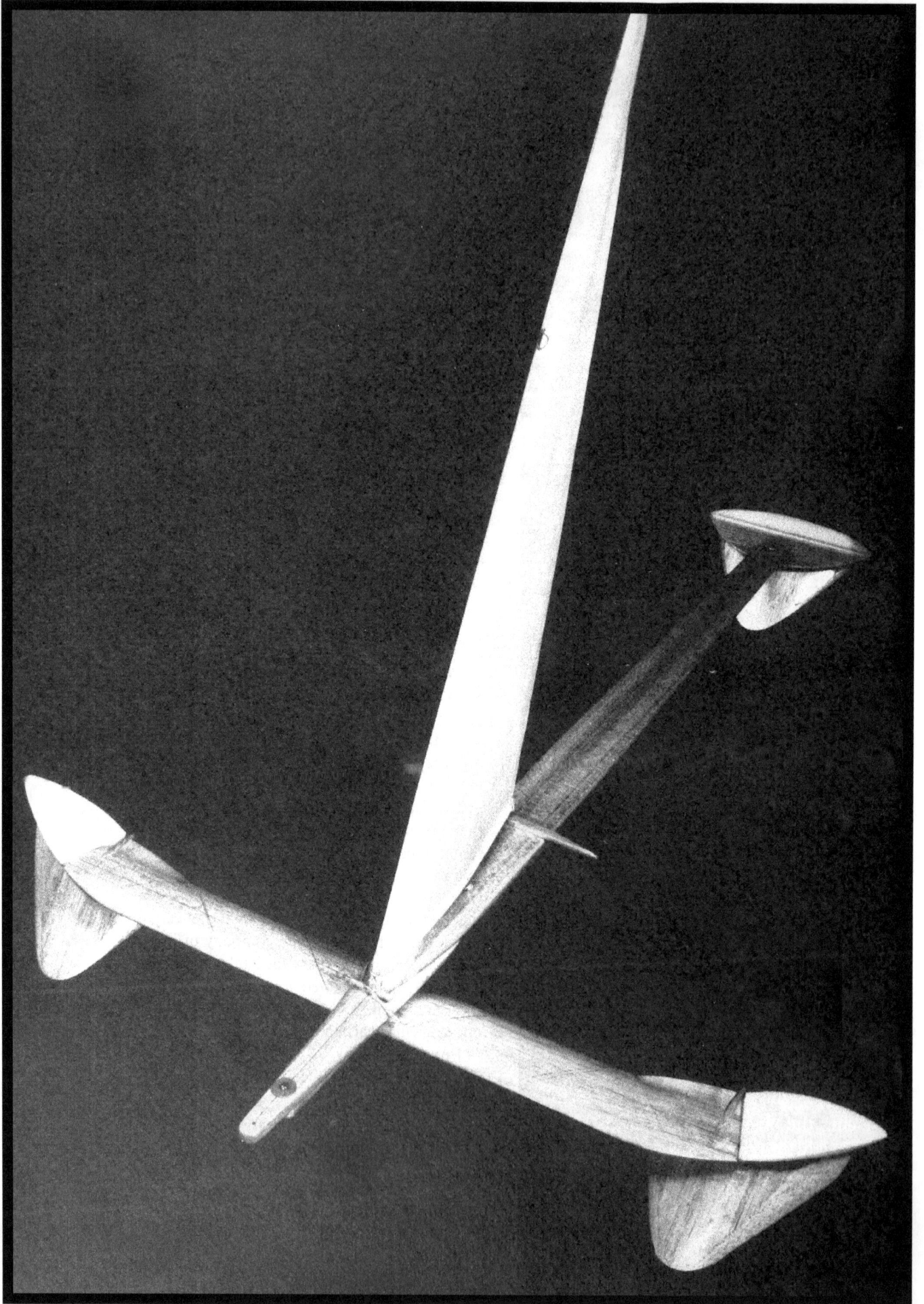

on a study of the ocean environment and investigated the potentials of many intriguing free-moving platforms for application to Naval needs. Of course the aerohydrofoil came in for its share of attention, but this one was too much my pet to be worked on during regular office hours only.

A New Synthesis

My study of the aerohydrofoil was conducted mainly after the ordinary day's work was done. I found; to my chagrin, that the extant literature contained little to guide me. All sources of information relating to this subject were uniformly meager. There was no comprehensive treatise on the aerodynamics of sails, no suitable analysis of the application of airfoils to sailboats, and no appropriate study on thick, tapered hydrofoils. Practically everything had to be thought out from the beginning and all of it together, otherwise I would risk falling into the net of minuscule thinking that enmeshed so many specialists. (The saddest commentary of all is that the best published analysis of the sail is in connection with its use in aircraft such as the paraglider and the sailwing. Unfortunately, this analysis was not adaptable to the sailboat and therefore had limited usefulness for me.)

My first computations showed quite clearly that potential speeds with hydrofoils could never be achieved to windward unless angles of attack considerably less than 15 degrees could be applied to the sail. This was patently impossible, because 15 degrees is just about the smallest angle of wind impingement essential to keeping a sail filled and properly curved. This problem was intimately tied to achieving the best lift-to-drag ratio for the sail, which the computations also showed to be critical when beating to windward. Available aerodynamic theory made it quite clear that the desired lift-to-drag ratios were associated with angles of attack of about five degrees.

Obviously, both needs could be met only with an airfoil. But what kind of airfoil could

be made lighter than a glider wing and yet capable of reversing its curvature to satisfy the requirements of beating to windward? After many false starts I came up with the scheme, partly visible in Figures 31 and 32, that accomplished a number of things at once with only a little more weight than an equivalent mast, boom, and sail. The scheme included:

1) true airfoil sections at all heights;
2) sweepback, for supplying directional stability with respect to the apparent wind;
3) curvature reversible with a sheet, as with an ordinary sail;
4) reverse angle of attack at the tip, to generate a restoring moment proportional to the wind's force.

Search for A Model Basin

My next great need was a model basin. I could build two- and three-foot models of balsa wood, foam plastic, glass cloth, and dacron in a matter of a few days; yet, with water all around me I had the utmost difficulty in finding a suitable test site. The small ponds of the region were surrounded by wind obstructions or swampy approaches. On one pond my model was attacked by a domestic goose who made it plain before witnesses that I was not to return. I feared trying the models on the ocean, because if they were successful I might lose them forever. It was too soon to attempt a transatlantic speed run. But one day I looked over a neighbor's hedge and beheld a fine, shallow pond, open to ocean winds. It was just right. The aproaches were low, and the grass around the edges was well-tended. Moreover, the banks looked firm, and, as it turned out, the water was of the right depth—deep enough for the models, yet shallow enough for retrieving runaways that became enmeshed in the water lilies.

So with a model boat under my arm, off I went to call on the Munginis one bright and windy Sunday afternoon, to request the use of their pond. Mrs. Mungini was quite interested

FIGURE 32

Symmetrical-Asymmetrical Aerohydrofoil

(Coming About)

68

in the enterprise; but as we talked, I found myself becoming even more interested in the trials and tribulations of Newport oceanside life. She had many intriguing anecdotes to relate, involving a number of community notables whom I had previously met. But I knew I was allowing myself to be distracted from the main purpose, so I tried walking toward the pond as we talked. Two hours later I had not yet reached the pond; the wind had died down and the sun was ready to set. No model sailed that day, but the pond was mine for the remainder of my stay at Newport.

Dramatic Windward Performance

The new airfoil was applied to a few symmetrical hydrofoil models. Behavior on the reaches was fast and stable, but the windward performance was as disappointing as ever. The models simply would not lift properly when close-hauled. I could contrive no symmetrical arrangement of hydrofoils to overcome the difficulties. However, when I hinged the afterfoils on the last model to allow the same or parallel inclination for all foils, the change in windward behavior was dramatic. The foils rose high in the water immediately, giving the model a decided increase in speed. This was so convincing that it put an end to all further experiments with strictly symmetrical boats. The Marianas Islanders knew the limitations of symmetry for a sailboat hundreds of years ago, but I had to learn them the hard way.

Back I went to studying new frame and foil arrangements. The first obvious step was to apply the new airfoil to the frame and hydrofoils of the asymmetrical model in Figure 27, which had done as well as could be expected with an ordinary sail. It was an impossible task. I could find no arrangement that retained the features of the new airfoil without introducing ungainly controls and engineering problems that would plague a man-carrying version out of existence. All the beautiful sim-

plicity of the early proa-like models was lost in the tangle of joints and adjustments.

I turned to a more fundamental approach to the problem and studied the degrees of freedom essential to coming about. From this I determined that a minimum of three joints was required, including the boom and rudder joints. Setting this as my goal, I methodically went through every possible combination of three hydrofoils fixed to two spars carrying one airfoil, all of which were constrained to rotate about three joints. At last I came to the arrangement shown in Figures 31 and 32. It had almost all the properties I sought.

Foremost was the alignment of all components in the direction of least drag. (Air drag, in particular, becomes quite critical in accomplishing high-speed windward motion.) In the last arrangement the principal member, which I call the fuselage, lies close to the line of the relative wind and presents the smallest cross section. This can be seen in Figure 31. An observer on the fuselage could easily delude himself that the aerohydrofoil was behaving just like an ordinary sailboat every time it came about. He would always feel the wind coming from ahead and see the fuselage head into the wind whenever the boat changed tacks. Unless he noticed that the direction of motion in the water had reversed for both the crosspiece and the rudder foil, he would be unaware that the boat was really duplicating the windward tactics of a double-ended proa.

The front leeward foil and the windward (middle) foil were rigidly attached to the crosspiece, so that all three were in line. Thus, the crosspiece pointed in the direction of motion through the water. Under these conditions the crosspiece could be brought very close to the water's surface, which was advantageous from the standpoints of center of gravity, buoyant support for overloads, and low aerodynamic drag. The front foil would, of course, become the windward foil when the tack was changed. When the crosspiece was adjusted for a tack, it became a streamlined outrigger that supplied some lift. The application point of this aerodynamic lift served also to develop, in proportion to the wind's strength, resistances to overturning and to downward pitching moments generated by the wind.

Trial—and Error

The first models constructed along these lines were far from perfect. Without the aid of wind and water tunnels to ascertain the precise centers of effort for air and water foils, the geometric estimates contained large errors. And since the parameters in any one plane were always dependent on those in all other principal planes, everything had to be right the first time before a model could perform properly.

Small changes in the position and direction of the resultant from the airfoil could cause either the forward or rear foil to submarine before top speed was reached Small changes in the distance of either foil from the fuselage could give the same improper behavior. Which was which? The same considerations complicated the problems of directional stability and heeling stability. More than once I was sure my analysis had identified the necessary change, and more than once I invited someone to take pictures of the corrected model's fabulous behavior, only to be embarrassed and disappointed.

Colonel Jack Wade, USMC, an associate at the Naval War College, bore the brunt of the false alarms. I must say that his patience was most helpful in these situations. He took time from his busy Sunday routine to bring a motion-picture camera to the model pond whenever I called. Only rarely did he see rudiments of the behavior I predicted each time the models were taken out for trials. But he did see the evolving promise in the concept, and he never failed to encourage me.

My tour at the War College and my work at the Institute of Naval Studies was rapidly coming to a close now, and once again I was faced with an interruption of my experiments with the aerohydrofoil. This led to another crash effort, at the expense of all but my most essential obligations, which paid off a week before I left Newport for Washington, D.C., my new duty point.

One Sunday morning I at last had the aerohydrofoil trimmed properly, and my model proudly beat to windward at 50 degrees off the wind. It moved as fast as the wind, when the wind blew steadily at 13 knots. It was as stable as a rock. Although according to theory it should have been capable of this speed at about 35 degrees off the wind, I knew it was now a matter of aerodynamic and hydrodynamic cleanup to attain the added performance. Most satisfying of all was the assurance that my calculation of the critical wind speed, about 13 knots, was quite correct. There was a pronounced improvement in boat speed at wind speeds above this value, especially on a broad reach, where the little model fairly flew. Needless to say, this happened the only day Jack Wade was unable to come with his camera; and no repeat performance was feasible in the short time remaining for me to pack up and leave.

Man at The Brink

Again I was a man at the brink, this time in an environment less favorable to aerohydrofoil experiments than China Lake or Newport. I was back in harness as Chief Engineer, at the heart of the Navy's vast weapons research and development establishment, with little time to do more than wonder about the day when I would ride the first full-scale version of the last model—still unaware of the drastic changes I was to make in the design of the aerohydrofoil. The sabbatical was over; I was thrust into the most exciting ulcer factory in the world, and I soon found my colleagues most envious of me for having picked up more unsolvable problems in a few months than all previous Chief Engineers of the Bureau of Naval Weapons had collected since the position was created.

FIGURE 33

Little Merrimac

FRONT VIEW

CHAPTER VI

The Washington Inventions

The last Newport model displayed many of the performance properties predicted for the aerohydrofoil. The design seemed to represent a reasonably good compromise of all the critical factors, yet I had an uneasy feeling that I had overlooked something fundamental. I knew that this model, like all the others, required ballast for stability, and therefore it could withstand winds only up to a certain strength before being upset. However, I was prepared to accept this as a fact of life. The aerohydrofoil, like other sailboats, would simply have to shorten sail whenever the wind became too strong for comfort.

Not being close enough to water to undertake construction of a man-carrying version, I did the next best thing, which was to put my thoughts down on paper. This coincided with an invitation by Donald Myrus, an editor, to prepare a book on the subject of fast sailboats. The period of reflection created by the need to record my ideas in preparation for publication, was exactly what was required for a last, fundamental look into all the problems. Probably it saved still another step in full-scale aerohydrofoil construction.

In reviewing the design for the last aerohydrofoil, I saw that up to a point it contained reasonable provisions for counteracting overturning or pitching moments and that some of the provisions were adequately proportional to the wind strength. But I had really done nothing to prevent the generation of overturning moments at the *source*. The resultant from the airfoil still had a large moment arm about the center of lateral resistance (C.L.R.) produced by the hydrofoils. It then occurred to me that I had been looking at the C.L.R. in only two dimensions, just as with ordinary sailboats, which simply was not my usual way of doing things.

A simple force diagram showed that if the C.L.R. were centered near the windward hydrofoil, and if the airfoil were tilted to wind-

72

ward, the airfoil resultant could be made to pass through the C.L.R. and the overturning moment would disappear. This meant that the leeward foils should provide an absolute minimum of leeway resistance and should be inclined for maximum lift; heretofore they had borne the lion's share of drift resistance.

Re-Analysis—Perfect Stability

A complete re-analysis was called for. The results led to the design in Figures 33 and 34, which contains all the experimental modifications to be discussed in the next few pages. To simplify the problems of construction, this model was improvised solely for starboard tacking. My immediate objective was to test a theory; the mechanism essential to coming about could await the outcome of the test.

The quest for a suitable model basin followed the same disheartening pattern experienced at Newport. The Reflection Pool, between the Washington Monument and the Lincoln Memorial, came closest to satisfying my needs. However, it was thirteen miles from my home in Fairfax County, Virginia—a great inconvenience for trimming and balancing the model. I knew from experience that there was little possibility of accurately computing the force vectors and their points of application in advance for the complex collection of surfaces on this model.

The small ponds near my home were blocked from the wind, either by houses or woods. Nonetheless, by exercising patience, I was able to employ the sporadic streams of air that filtered through the trees surrounding a neighboring lily pond. Despite the turbulence, I could discern the general direction of the wind with cigarette smoke, and, at the expense of a parched mouth and a carton of cigarettes, I learned enough from the reactions of the model to make the first gross adjustments.

Position of Air Rudder

I found, first of all, that the position originally planned for the air rudder, at the upper tip of the swept airfoil (see Figure 28),

was not suitable. Because the airfoil was inclined to windward, the drag induced by actuating the rudder created a turning moment that partially cancelled the desired directional control. As a result, the model had excessive weather helm. No matter how far over the rudder was turned, the model became stabilized with the wind on the wrong side of the airfoil, giving a negative angle of attack on both airfoils and hydrofoils. The model was perfectly stable going backwards!

The correction was easily made. As shown in Figure 34, the rudder was placed over the rear hydrofoil, a position of greater mechanical advantage. Moreover, both drag and normal force would now contribute to turning off the wind. The shift in the rudder's position, however, eliminated the last vestige of an aerodynamic restoring moment. After this change, the model could not possibly live in any but the lightest breezes unless it did, indeed, prove to have little or no overturning moment. In this respect the arrangement was a good one for testing the efficacy of the new principle. The amount of ballast needed to keep the model upright would give me the answer quite directly.

The little model was more than a match for the thick boundary layer and high turbulence that existed over the test pond. It responded easily and quickly to all the vagaries of the wind and became stabilized to a proper angle of attack as soon as the wind blew steadily in one direction. The tendency to overturn was less than for any previous model and was easily countered with a small amount of ballast.

The wind strength at the surface of the pond, however, was rarely more than three knots, and even though the model rapidly accelerated to almost the same speed, the motion was not great enough to produce significant hydrodynamic lift. Furthermore, at these speeds the model was in the regime of maximum wave drag for its scale. (The airfoil was overweight, which caused the fuselage to contact the water and increased the velocity of maximum wave drag from about one knot to about 2½ knots.) A steady wind, closer to ten

FIGURE 34

Little Merrimac

SIDE VIEW

knots, was required for a decent test, and obviously a more suitable model basin was needed.

A Christening

While I ruminated over this problem a distinctly different matter called for my attention. I had named none of my previous boats, and my new model clamored for recognition. Gordon Baker of Minnesota, who invented the first working hydrofoil sailboat, named his craft the *Monitor* apparently in deference to Northern tradition. Now that I lived in Virginia and had assumed the role of a Southern gentleman, I certainly could not ignore this challenge. The South's answer would, of course, be *Merrimac* (the original name of the *Virginia*, which met the *Monitor* in combat).

With the question of regional pride settled, I attacked the problem of the model basin with renewed vigor. Believing the *Little Merrimac* ready for a speed run, I took it to Reflection Pool on a day when the wind was blowing partly in the long direction of the pool. Fram Ellis, an aerodynamicist at the Bureau of Naval Weapons, came along to prevent the model from smashing itself on the stone banks. Much to our dismay we found the wind turbulent and variable from minute to minute. However, we did find a corner with some wind constancy—enough to learn that the *Little Merrimac* was at the limit of its control sensitivity. The air rudder had to be turned so far over to keep the boat on course that the reverse force coming from the rudder cancelled a very large part of the driving force. The model was overweight to begin with, and with this serious loss in driving force, its fuselage could not break contact with the water. Unless the wind blew considerably stronger than the three knots we found at the pool, the model would not be able to rise on its hydrofoils. Obviously some modifications were needed, and a model basin closer to home was imperative for testing the modifications.

The Melpar Company came to the rescue. Its plant, located in Falls Church, Virginia, had a small, decorative pond landscaped into the adjacent grounds, and the president of the company, E. M. Bostick, kindly granted me the privilege of using it as my own private model basin. This was a great convenience because the pond was only a few miles from my home and the "No Trespassing" signs all around it would guarantee privacy. The wind turbulence left much to be desired, but the pond was adequate for cursory testing of the changes made on the model.

True Aerohydrofoil Performance

After removing all the weight I could from the leeward side, and after increasing the moment arm of the air rudder as much as the dimensions of the model would permit, I tried it on the pond. Grave doubts entered my mind as I watched the *Little Merrimac* turning and twisting under the influence of changing gusts. For half an hour I wondered if she would ever climb on her foils so that I could finish this book! Suddenly a sustained stream of air found her in just the right orientation, and away she scampered. I sighed with relief as I watched her perform. All the principles seemed to work. With a few more runs I was able to determine that the wind occasionally reached about four knots. At that wind speed the boat's fuselage left the water surface, whereupon she gave true aerohydrofoil performance—she beat upwind at about 40 degrees off the wind at almost the speed of the wind. She displayed practically no heeling tendencies, and she required little rudder for directional stability. Riding on her foils alone, she was perfectly level. The *Little Merrimac* was now ready for a high-speed run.

Three more attempts were made to test her on Reflection Pool—all dismal failures. Between the turbulence produced by the Lincoln Memorial and the thick boundary layers underlying the erratic gusts, the *Little Merrimac* did not have a chance. The last attempt was particularly disastrous. One gust slapped her on the wrong side and pushed her weather hydrofoil down so deeply into the water I thought it would never come up. A second vicious blow caught her from the other side just as she was springing back. Coupled with the boat's own restoring gyration, it was enough to flip her on her side. My poor model had to drift the entire length of the pond in this humiliating condition before she could be retrieved. Both airfoil and rudder, which were not protected against wetting, suffered severely from this treatment, which motivated me to resolve never to test another model on Reflection Pool.

At the home of a kind neighbor, Mrs. Grace Summers, whose lily pond I had used for my earliest model tests in Virginia, I at last found an answer to my quest for a suitable test site. Her son, Bill, had been following my experiments with some interest, and, noting the inadequacies of the lily pond, he suggested that I investigate Lake Accotink in Fairfax County. Upon visiting the lake I saw immediately that it would be almost ideal. It was large enough to give the wind a clean sweep, and it was conveniently close—only 15 minutes away. Best of all, it had public rowboats for hire. In a rowboat I could start the *Little Merrimac* in the middle of the lake, away from the turbulent winds near shore; and I could row out to retrieve her if she ran into trouble.

Built-in Resistance to Capsizing

Recalling what had happened in the last test on the Reflection Pool, I modified the windward foil before trying the model on Lake Accotink. I calculated that if the lower part of this foil were inclined 35 degrees away from the vertical, instead of 20 degrees, it would develop twice the upward force if the wind should strike the model from the wrong side. This would prevent the foil from dipping so deeply in a strong reverse wind. Conversely, under a strong gust from the proper side, the foil would offer more resistance to being pulled out of the water than it previously did. I had some misgivings about this change because I knew it would reduce the efficiency of the model somewhat, but, as it turned out, the loss was not serious.

One chilly Saturday afternoon late in October, 1962, I gathered my model, my wife, and my courage, and rode to a fateful test on Lake Accotink. When we arrived at the lake, the wind, which had been good until then, began to die, in conformance with the usual conspiracy. By the time I rowed out some distance from the shore, however, it had returned in half-hearted gusts. I judged the boundary layer to be small, which was good because it provided a wind velocity at the rudder level commensurate with the flow over the airfoil. (A thick boundary layer, like that over Reflection Pool, simply would not give the rudder all the correcting force it needed safely to turn the boat into the wind when pressed by a gust.)

With my fingers crossed I placed the *Little Merrimac* on the water and waited, with my wife, for a favorable gust. It moved slowly away from us, changed direction a few times, and then, as the water ruffled before it in a ten-knot wind, it rose on its foils to fly along as fast as the wind at less than 40 degrees off the wind. These estimates were based on years of experience in feeling such winds on my face and watching my models at such speeds. If human error was made, it was on the side of overestimating the wind speed and underestimating the boat speed and the angle off the wind, because what I saw seemed much too good to be true. *Little Merrimac's* performance was even better than my theoretical calculations had predicted for her battered surfaces. Neither the aerodynamic surfaces nor the hydrodynamic surfaces could have developed a net lift-to-drag ratio as great as ten, after all the changes were made in the *Little Merrimac*, yet it matched the behavior anticipated for an aerohydrofoil with such a ratio.

Theory Proven

A few more runs verified the first results. In fear of pressing my luck too much I retrieved the *Little Merrimac* and headed for home, elated and contented. On the ride back I reflected over the notions I had entertained three years earlier, and how vastly the original concepts of the aerohydrofoil had changed. How often I had wondered whether I was pursuing a will-o'-the-wisp. At China Lake I originally sought only to eliminate the hull, but after doing so I found that it was also essential to eliminate the sail. In Newport I eliminated the sail and found later, in Washington, that this was not sufficient either. But now, at last, I had removed the ballast, and with this accomplishment came a satisfying, intuitive feeling that the last obstacle to high-speed sailing was demolished.

The final part of my story lies in the remaining chapters, which are more orderly. In them I recapitulate my technical findings over the last three years. I have tried to arrange the findings along lines that follow normal processes of instruction in a new field: elementary and simple principles first, followed by gradually more complex combinations, until every purpose for every part needed for a manned aerohydrofoil is revealed.

PART THREE

Technical Summary

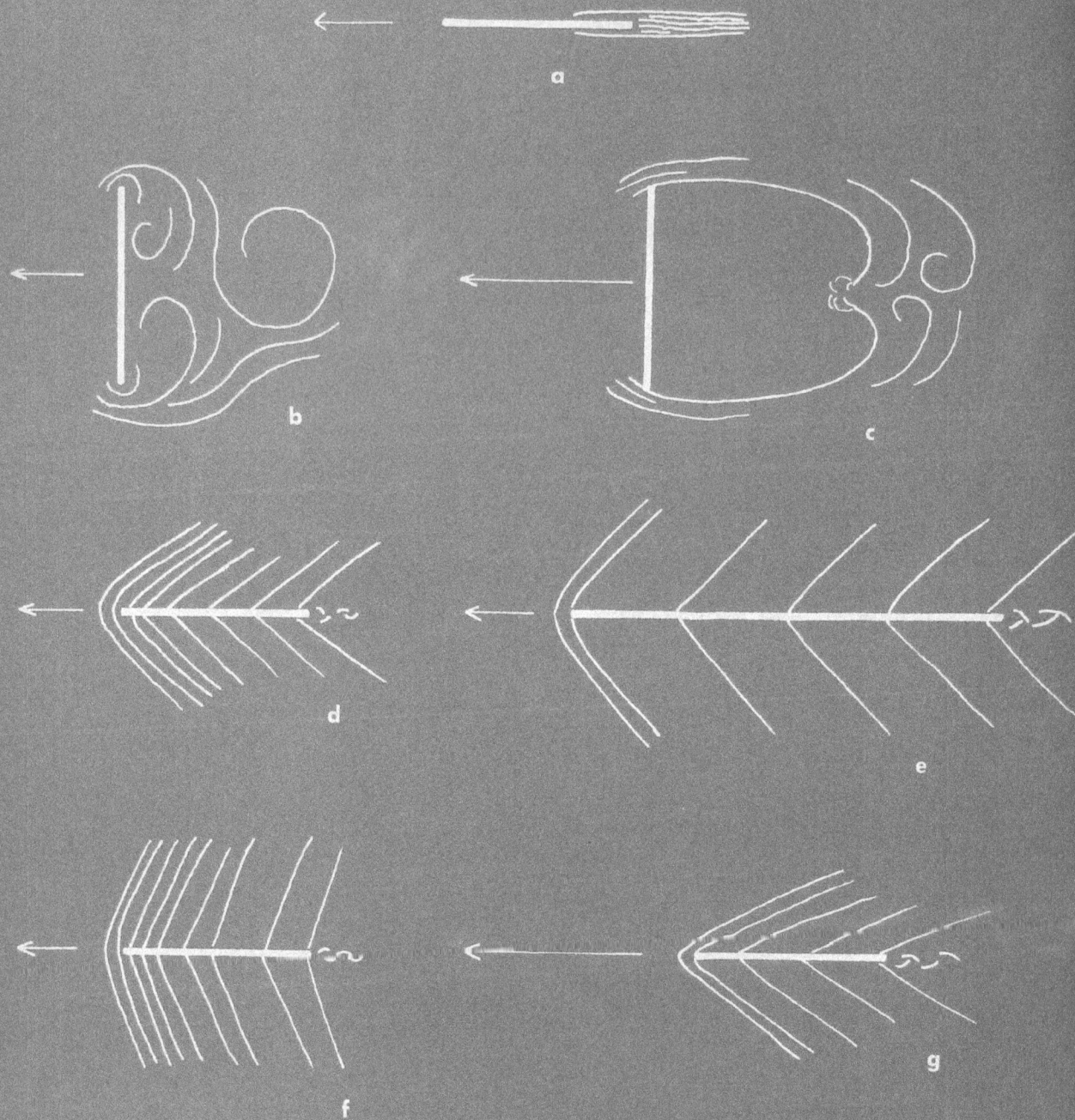

FIGURE 35

Resistance to Motion

The simplest kind of resistance is illustrated in (a). A thin blade drawn edge-first through the water is retarded almost solely by skin friction introduced at the sides of the blade. Two forms of inertial drag are shown in (b) and (c). A flat plate, drawn through the water at increasing speed, would first develop turbulence, and ultimately, if forced to move at sufficiently high speed, it would leave a cavity behind it. Thin blades moving edge-first generate waves that increase in size with increasing waterline length (d,e) or increasing speed (f,g). The increased wave size corresponds to increased wave drag. With higher speed the chevron of waves becomes sharper. Beyond critical speed values the waves diminish until the primary source of energy absorption reverts to case (a), (b), or (c).

CHAPTER VII

The First Obstacle
(A Short Course in Hydrodynamics)

In approaching a problem it is possible to exercise either of two options: solve it or eliminate it. There is still a third way, which is to ignore it and hope that the problem will disappear. The last, unhappily, will not work for the sailboat speed barrier. I have tried it for about thirty years with absolutely no luck. Nor has the orthodox method of working vigorously on the problem with standard tools been much better. The last course left is the oblique one—the one that resembles the process of beating to windward. One does not solve the problem of sailing directly into the wind; the only way to make progress upwind is to sail in some other direction. This aphorism also holds for the attack on the sailboat speed barrier. To make a big step forward one must avoid the direct approach, which is handicapped by the inherent limitations characteristic of current sailboat design.

The first obstacle to the achievement of high-speed sailing is the hull itself. Refinement of the hull's lines is not the solution, however. The reason has been known for over one hundred years. To better understand it we should review the factors that contribute to the resistance of a boat's motion through the water.

The drag of a surface vessel is sometimes treated in three parts: viscous drag, more picturesquely described as skin friction; inertial drag, which is a measure of the water's reluctance to move aside and close behind a moving body; and wave drag, the most visible part of the protest made by the water against trespassing.

The first two are inextricably interwoven for most practical surface craft. However, simple experiments can be contrived to help us distinguish between them (Figure 35). A thin blade passed edge-first and vertically through water very nearly illustrates the action of pure dynamic viscosity. The same blade passed broadside through the water encounters almost pure inertial drag. Depending on the speed of

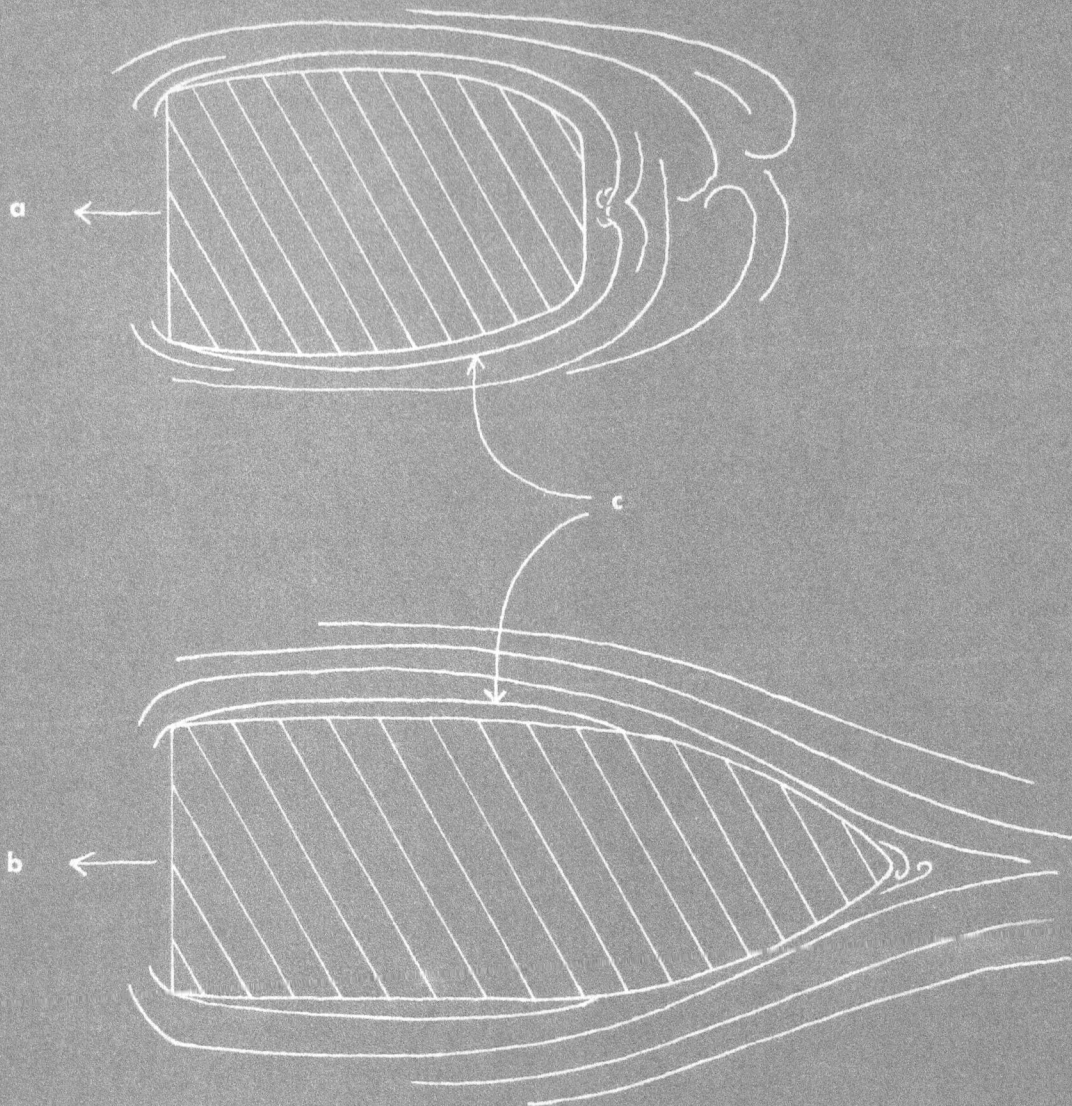

FIGURE 36

Streamlining

A relatively large body can be placed in the cavity (c), produced by a flat plate, without increasing the resistance to forward motion. The lower diagram shows how to achieve a reduction in total drag by enlarging the body. The closing walls of the cavity can be forced to react against the streamlined tail, developing a forward driving component.

the blade and other conditions of the experiment, substantial wave drag may have been introduced each time. This can be determined by simple inspection of the water surface, provided the waves are not confused with the wake. The wake, of course, is revealed by turbulence, or the churning of water caused by both viscous and inertial drag. If these experiments are conducted in a quiet basin or bathtub, it will be noted that the waves are more easily observed when the knife is moved edge first and that the turbulence seems to hide well-behaved waves when the blade is moved broadside.

Drag Proportional to Area

If such experiments were continued, using blades of equal thickness and length but of greatly differing width, it would be found, through muscle reaction alone, that, at least for constant velocity, both viscous and inertial drag are apparently proportional to the area of the blade. It would also be found that as long as the flat face of the blade was parallel to the direction of motion, giving the blade sweepback would make no appreciable difference in the viscous drag. Nor is there any advantage in drawing the blade tip first through the water, so long as the wetted area remains the same and the blade is not rotated about its long axis. With similar simple experiments, one could get some idea of how blade roughness would affect viscous drag. But this can be guessed without going to the trouble of roughening and ruining good knives. Roughness increases skin friction, as we all know from other experiences.

When the broadside knife is pulled very fast, a cavity is formed in the water behind it. Logically enough, this phenomenon is called "cavitation," and for the case in point it is named "ventilation" because the hole is open to the atmosphere, thus changing the conditions somewhat. But in either case it constitutes a visible display of the water's inertia, which resists the change in direction essential to closing behind the body. The cavity can be made to extend down to the tip of the knife if the speed is increased. Why doesn't the cavity form along the whole blade at once? Because the higher pressure lower down in the water helps to push the stream back around the blade. From this it is concluded that external pressure has something to do with flow lines.

The proportionality of inertial drag to area is not as clear as is that of viscous drag to area. By virtue of the cavity, the wetted area has been cut almost in half; yet the smooth increase in force required to achieve cavitating speeds has not been affected appreciably. At this point, one might conclude that muscles are not sufficiently sensitive to detect the difference. A little study of the cavity could lead to another conclusion, however. Suppose, in imagination, we were to add more "boat" to the broadside knife and fill in behind it with a surface that stays exactly within the bounds of the cavity (as in Figure 36a). We have increased the volume and the total area without increasing the wetted area, and we have not changed the cavity or the front surface in contact with the water. Therefore, we have not changed either the viscous or the inertial drag. As long as we stay inside the cavity we could make the shape of the added "boat" anything we liked, without changing the total drag. Only the frontal area has remained measurably constant. One can infer, therefore, that inertial drag depends on the frontal area, or the cross-sectional area projected in the direction of motion.

Bathtub Experiments

At the risk of getting into complexity too soon, this conclusion can be checked by inclining the blade from the vertical without turning either edge in the direction of motion and moving it at speeds that do not cause cavitation. Drag changes will develop in proportion to the altered projected, cross-sectional areas, although new vertical forces will appear to

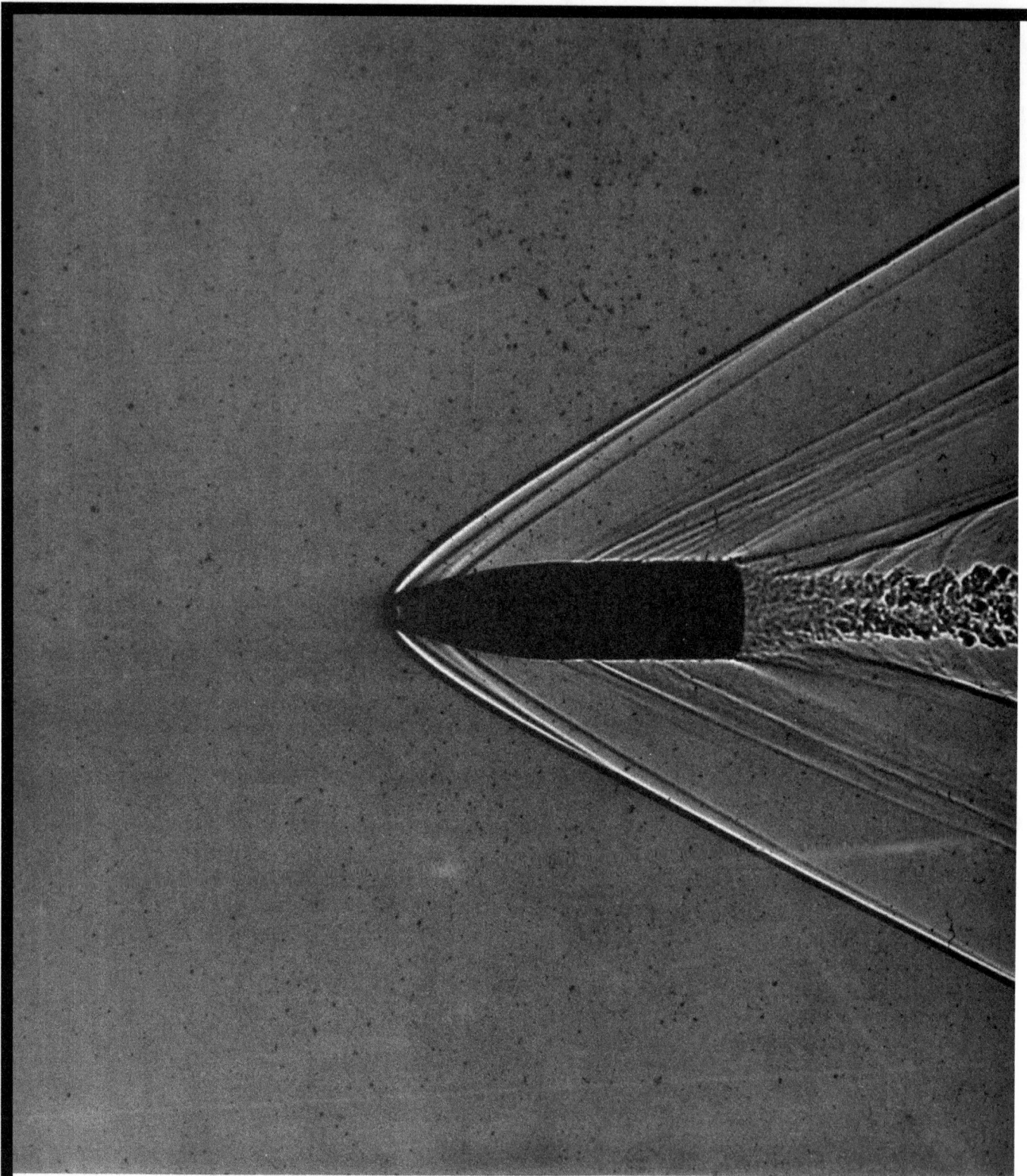

FIGURE 37

Wave Analogues

This Schlieren photograph of a projectile moving faster than sound in air shows how closely the waves and the wake resemble those of a boat moving through water. The formation of waves in any medium is frequently caused by a body moving through that medium at a velocity higher than the natural velocity of wave propagation. The phenomenon is called Cerenkov radiation. An airplane moving at or above the speed of sound develops a shock wave; a surface vessel moving at speeds greater than its own characteristic wave speed develops water waves; an electron entering

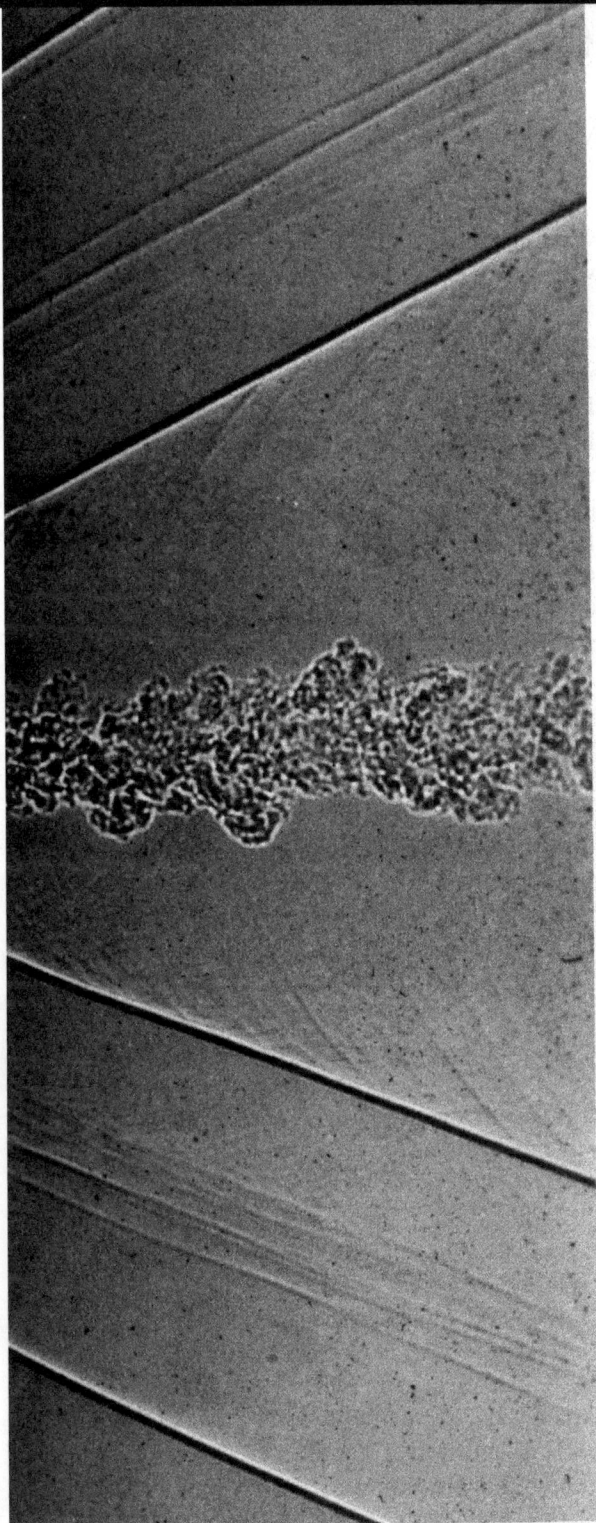

confuse the experiment. But nothing yet challenges the contention of proportionality between projected area and drag. It is especially confirmed when we note that the wetted area has not been decreased in the process of changing the projected area.

Unfortunately, nothing is ever simple in the water. To demonstrate that the rule stated above has exceptions, it is necessary only to increase the speed enough to create a cavity wider than the knife blade and to continue the "thought experiment" a little further. Inside this new cavity we could place a body of larger projected area without increasing the drag. But to consider this relationship in any greater detail now would take us too far afield.

Suppose the imagined boatlike extension of the knife were made long enough to trail in the water beyond the cavity, as in Figure 36b. If it were faired properly, we would find that despite the increased wetted area the total drag had decreased. How do we explain this? As a matter of fact, it was predictable. A streamlined tail would have inclined surfaces, like a cone. Transverse forces applied to it must develop a net forward force. Where do the transverse forces come from? They come from the walls of the closing water cavity. They squeeze on the tail and push it forward, just as our wet fingers would if we were to squeeze the inclined surfaces of a bar of soap. The water has been deceived into returning some of the energy put into it farther forward in the line of flow.

This example indicates that the problem of reducing drag is not always alleviated by reducing the wetted area. On the contrary, it is more generally true that a submerged body develops its lowest drag when it is carefully streamlined to run without creating a cavity.

Drag and Velocity

But precisely how is drag influenced by velocity? This cannot be found out easily from bathtub experiments. However, the informa-

a transparent substance at a velocity higher than the local velocity of light develops a light wave. The direction of the wave front is related to the direction and velocity of the moving object and to the local velocity of the radiation under consideration.

tion is readily obtained from a conventional study of fluid dynamics—more readily, perhaps, than the other points of information discussed in this chapter. The viscous drag of a surface increases in direct proportion to the velocity of the fluid flowing over it. The inertial drag, on the other hand, is proportional to the square of the velocity—obviously a faster rate of increase. A deeper study of fluid dynamics would reveal that viscous drag becomes proportionately less significant as body size and flow rate increase. Consequently, wetted area becomes less critical than cross-sectional area as boat sizes and speeds become greater. Accordingly, most of the larger and faster displacement hulls are relatively slender.

We come now to the part of a surface vessel's drag that is more difficult to comprehend —its wave drag. In her usual way, nature has played coy in an area that is far more critical to the design of surface craft than are the simpler sources of energy absorption we have been discussing. What she has done here must be seen to be believed; and so back to the bathtub.

Demonstration of Wave Drag

If the knife is drawn edge first very slowly through the water, no waves are produced. As speed is increased, the first waves form on a front that runs almost perpendicular to the direction of motion. With more speed the amplitude of the waves increases, and the waves change their angle to form a sharper V. Two kinds of waves, capillary and gravity, complicate the phenomenon, and they follow somewhat different laws, but we need not try to distinguish between them for our purposes. The appearance of the water waves is exactly like that of the shock waves produced by an airplane going through the sonic barrier (Figure 37)—and it is in fact the same sort of phenomenon.

If blades of different widths should be used in these experiments, it will be observed that the wider blades must be drawn through the water faster in order to generate easily discernible waves; and, once created, the waves are not as crowded. This fact suggests that waterline length may have something to do with wave form and wave velocity. But the best surprise is yet to come. As speed is increased still further, and the chevron of waves becomes very sharp, the knife seems to disturb the water less. Finally, we find that we can pass the knife through the water so fast that almost no waves are produced except at the beginning and end of the stroke. These phenomena, by the way, are fairly independent of the depth of immersion, provided the knife has the same width throughout its length and it is kept perpendicular.

All very interesting, but why the excitement? The inertial drag produced by the broadside knife was certainly much greater than the puny forces produced by these waves, which offer hardly any resistance. However, the broadside knife had no usable volume to speak of. If we were to design a respectable boat with the same frontal area, capable of carrying a box of breakfast cereal across a pond, its length would be much greater than the knife's width; in fact, it would be three to six times the *length* of an ordinary table knife blade, and we would find its wave drag quite a different matter. Moreover, as soon as the boat was made truly buoyant, the effect on the boat of its own induced waves would be impressive.

If one rides a boat in smooth water and looks over the side from a low position, he observes, at first, a number of low waves radiating at high angles from the hull, just as in the knife experiment; except that the wave lengths are much greater (Figure 38). As speed increases it seems to the observer that both amplitude and wave length increase also; for, although the first wave is fixed to the bow, the successively smaller waves behind it are spreading out toward the stern, and where there were four crests before, there are now three within the waterline length of the boat.

FIGURE 38

Wave Development

Sketch (a) illustrates the appearance of low-speed wave generation. When the hull bridges a large number of waves the wave drag is relatively low. Sketch (b) depicts the critical condition in which the second wave is exerting a forward force on the stern, in the manner of a wave pushing a surfboard. Higher speeds result in disproportionately large drag increases and ultimately lead to the situation shown in sketch (c), wherein the wave drag is maximized.

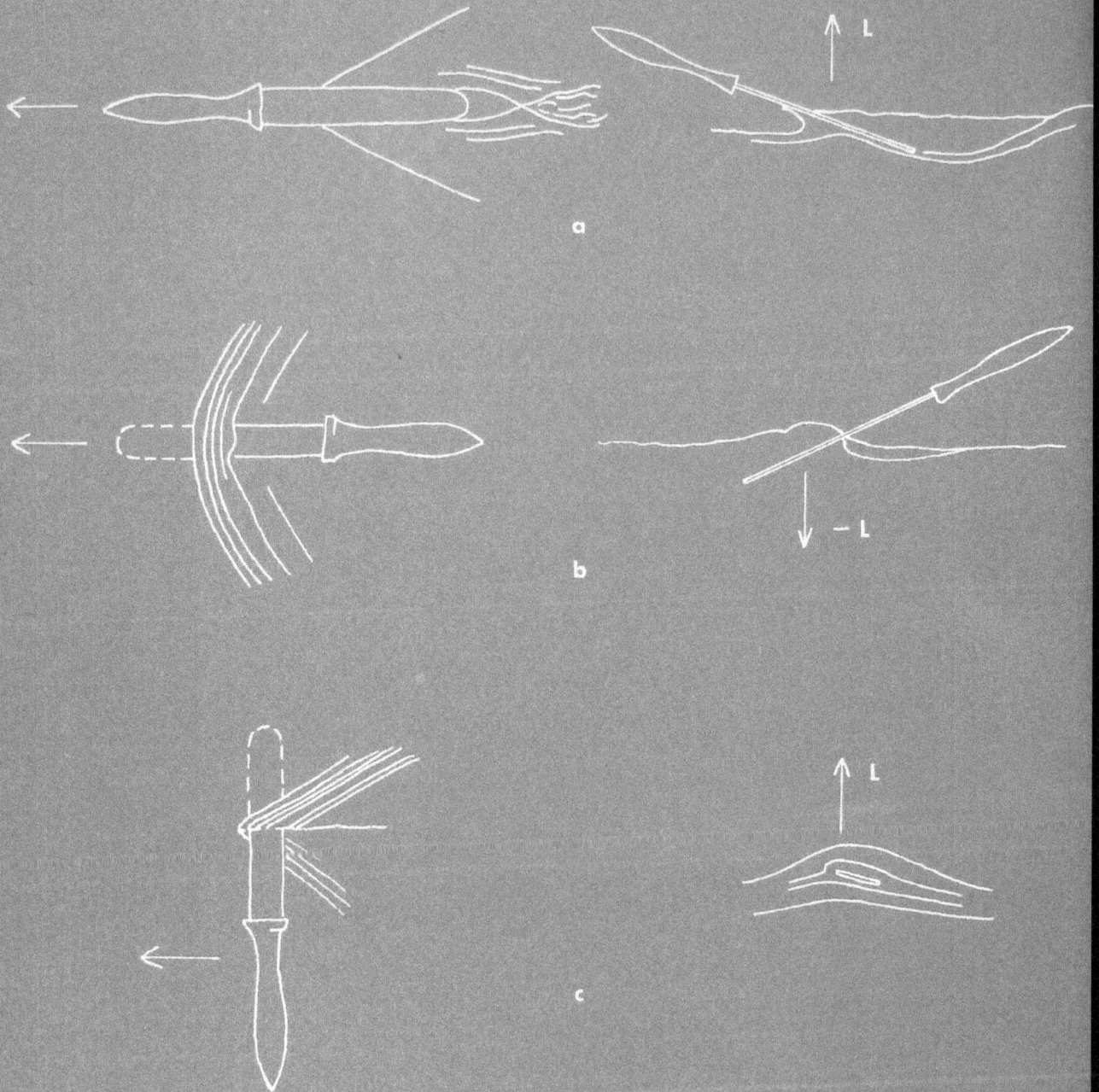

FIGURE 39

Vertical Hydrodynamic Forces

The diagrams at the left show plan views of a knife drawn through the water in various attitudes. The corresponding diagrams on the right are vertical sections showing the direction of lift and the reactions with the water. The illustrations correspond to (a) planing, (b) submarining, (c) hydrofoiling.

This is not all due to the fact that wave length increases with velocity; it is partly an illusion, because as the wave front sweeps backward more sharply with increased speed, the boat is really cutting across its own waves at smaller angles.

Absorption of Energy by Waves

As greater velocity is reached, the third wave breaks free of the stern, and an increased effort, over and above the previous rate of increase, is required of the power source in order to gain additional speed. But all is soon well again, in fact better than before, for the increase in power required to drive the vessel still faster is perhaps more modest than what was experienced before running into the first trouble. The respite is short lived. What takes place when the second wave begins to leave the stern is almost never witnessed in the behavior of full-scale displacement boats. No self-respecting marine architect would carry the propulsion requirements that far. However, through extrapolation of water-tunnel data, the results of attempts to drive the boat still faster may be depicted.

When the crest of the second wave backs off from the stern, not only do we find the power requirements mounting out of all proportion to the small increases in speed, but the very attitude of the boat suffers a marked change. As more speed is forced on the boat, the stern soon settles in the trough between the two waves and the bow reaches for the sky. It all looks like the behavior of a surfboard, only backward. The boat is going uphill. Where is the energy being dissipated? It is being carried away by the wave that is crashing dramatically on the shore and eroding the beach.

The energy absorbed by the waves is now so great that it may exceed the combined viscous and inertial drag. In fact, at this speed it would take considerably less power to propel a boat of the same displacement under the surface, despite the greater wetted area. For some boats, this progression can be carried further, to a point where the boat gradually moves up higher on the crest of its own wave and settles down to more level riding and better responses to additional power increases. But it is not found to be an economical practice for pure displacement boats.

How do we explain the ups and downs in drag as the waves successively leave the stern? What is happening here? If the stern is properly inclined, a wave that passes under it gives the boat a forward push, just as a wave pushes a surfboard. By designing the boat properly and running it at the right velocity, we can, as we did once before, fool the water into returning some of the energy we put into the wave at the bow. If the wave leaves the stern, however, the free ride is over.

Wave Drag: A Function of Waterline Length

As we might have suspected earlier, the speed at which the wave drag is greatest is a function of the waterline length. For well-designed, single-hulled, keeled sailboats, the velocity at which the second wave begins to leave the stern is given by the following formula:

Velocity in Knots =

$$1.6 \sqrt{\text{Water Line Length in Feet}}^{*}$$

This formula tells us that the best speed we can hope for in a 25-foot displacement boat is about 8 knots, or 9 miles per hour. In an 80-foot boat (America's Cup class) it would be about 14.5 knots, or 16 miles per hour. On this basis it would take a 400-foot sailboat to stand a chance of making 30 knots—and then only in a howling gale. The sea kicked up by such a wind would forbid the experiment.

Speed Limitations

There is really nothing we can do to mitigate this rule for keeled boats with all ballast in a fixed position. If we add more sail, we must add more ballast or deepen the weighted

---

* This relationship is numerically correct, but not dimensionally correct. It would scandalize the pure physicist, who knows that velocity cannot equal the square root of length.

keel. The drag increases about as fast as the driving force. We can beat the game a little by providing movable ballast (a nimble crew for light boats), or by widening the boat, which in effect does the same thing. Still more righting moment can be gained by splitting the boat and separating the two parts as widely as we dare. This gives us a catamaran or a proa. On such platforms the movable ballast or the fixed asymmetrical ballast can be of maximum effectiveness in countering the wind's overturning moment. Less ballast is needed because it can be placed at a greater distance from the axis of rotation. With the help of all these tricks, and under the right conditions, a 25-foot catamaran can make close to 20 knots. But the wind would have to be great, the water reasonably smooth, and the point of sailing a broad reach. Catamaran sailors often swear they were pointing up close to windward when such speeds were achieved. We will learn later how some of them came to be confused.

The flying proa discussed earlier no longer exists in its original form, but there is good basis for believing that one with hulls and a sail built in accordance with modern knowledge could trounce a catamaran. However, even a proa would ultimately be limited in the same way as a catamaran. Eventually the small windward outrigger lifts out of the water, and the pitching moment derived from the sail depresses the bow of the main hull, sometimes forcing it below the surface until the sheets are eased. Under these conditions it does no good to try to relieve the lee hull by adding more ballast to windward. This actually forces the lee hull down deeper into the water! What is needed now is a dynamic lifting force found neither in the catamaran nor in the proa.

An extension of the projected-area experiment in the bathtub provides a clue to the solution (Figure 39). As an inclined knife is drawn forward and backward, the perpendicular force reverses direction. Depending on the angle and the speed, it varies in strength. When the handle end is forward (Figure 39a) the blade lifts. When the speed is made sufficiently great, a wake forms that looks exactly like the "rooster tail" left by a planing boat. When the tip of the blade is forward (Figure 39b), the force is downward, and water tends to pile up in front as it does with a diving submarine. The dynamic downward force builds up very rapidly and seems much stronger than the lift force when the knife is planing, demonstrating what happens when the bow of a long, straight hull dips ever so slightly below the surface at high speed.

Demonstration of Hydrofoils

When the knife edge is turned at right angles to the direction of motion and worked like an airplane wing, the lift force is considerably greater than before. If the water surface is pierced at not too shallow an angle, and the blade is moderately inclined to the direction of flow, more than twice the lift possible with planing action is developed. The turbulence displayed by planing the knife has disappeared, telling us that less energy has been put into the water.

How are the lift and drag forces related? Why did the blade deliver more lift when we forced it to simulate an airplane wing (hydrofoil)? Simple explanations are often disarming and dangerous. But some elementary considerations can be introduced now without misleading anyone. A planing boat has only one surface with which to develop lift—a high-pressure one under the hull. A hydrofoil has two, a lower surface that works like a planing boat, and a low-pressure upper surface that is much more effective. The upper one can more than double the lift. Careful experiments indicate that under ideal conditions a planing hull can develop a lift force about six times the drag force. This may sound impressive until we learn that the lift from surface-piercing hydrofoils can be 14 times the drag value, and for completely submerged foils the ratio can be still higher.

Obviously, if it is lift we want, more can be obtained by resorting to the hydrofoil sailboat than to the planing sailboat. The planing boat is poorer than the best displacement boats until it reaches planing speeds. When sailing enthusiasts describe the phenomenal increase in speed obtained so abruptly at this point, they are really only talking about a few knots. What they are expressing is gratitude for small favors. There is nothing inherent in the planing sailboat that will improve stability or seaworthiness. In fact, to get the full advantage of the planing surface, it is necessary to bring the movable ballast farther out to windward than ever before. The flatter stern associated with the planing hull, needless to say, does nothing to help take a pounding sea.

A Hull-less Sailboat

And so we are left with the last possibility—the application of hydrofoils. Unfortunately, until quite recently efforts to apply hydrofoils to sailboat hulls have resembled the application of hydrofoils to motor boat hulls. Until the foils move at respectable speeds, they do little or nothing to improve the stability problem or the wave drag situation that comes free-of-charge with the hull. Conditions are much better when the velocity is high enough to enable the foils to lift the hull out of the water. However, if an ordinary sailboat achieved a speed of 30 or 40 knots with the help of foils, which seems possible, any impact of its hull with a heaped sea could, with equal possibility, produce an indescribable catastrophe. Removing the hull would eliminate the problem of the transition from buoyant support to dynamic support and also diminish the hazard of an impact with green water.

But removing the hull is not the entire answer. When speeds become greater than the wind, a factor that we have not yet examined in detail begins to loom in importance. In fact, if not overcome, this obstacle makes the high-speed sailboat hardly more than an academic exercise. Until it is resolved there is no real advance in the rate at which we can make progress to windward. The arrangements of sails, hull, and hydrofoils used in the past have not yielded a real improvement on this point of sailing. We shall see why when we delve into the nature of this obstacle.

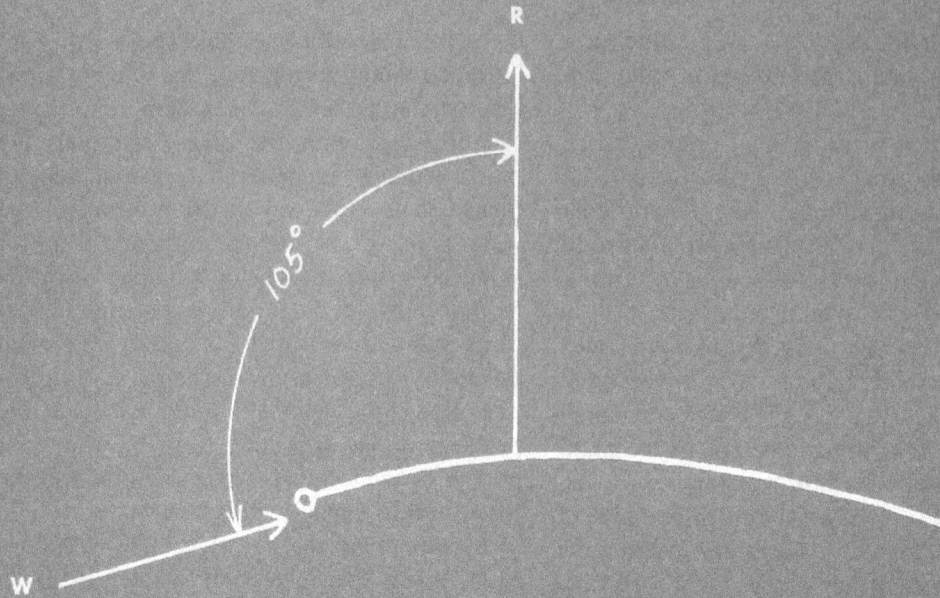

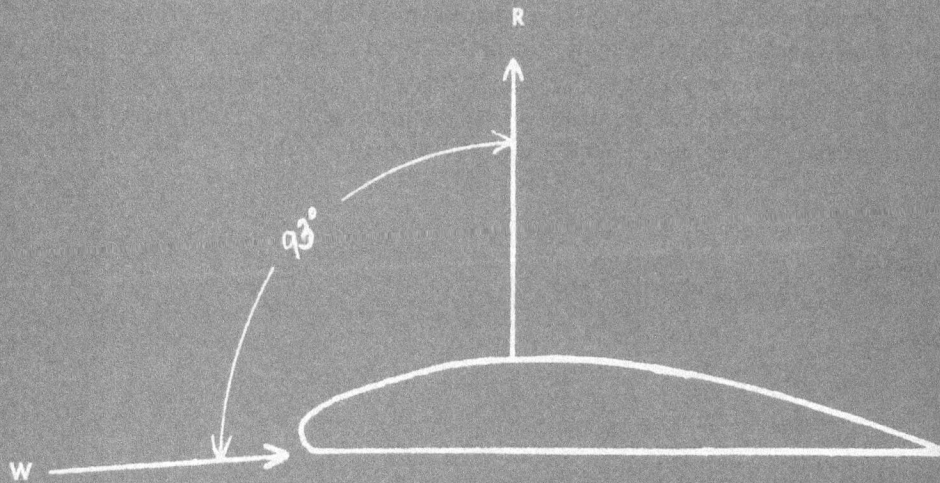

FIGURE 40

Angle of Attack

Because a sail (top diagram) cannot be filled properly at low angles off the wind, the angle between the direction of the wind (W) and the direction of the resultant force (R) developed by the sail cannot be less than 105 degrees. For an airfoil (bottom diagram) the angle can be less than 93 degrees.

Chapter VIII

The Second Obstacle
(A Short Course in Aerodynamics)

Strictly speaking, a sailboat does not have to be an efficient energy machine. With power galore, literally scattered to the four winds, one need feel no remorse about squandering it. Misuse of the wind does not diminish a sailboat's range, in contrast to the result of wasting fuel in a motorized vessel. Despite all measures to husband this free source of power, most of it will go to waste anyway. If the sails are inefficient, it is necessary only to crack on more of them. All that is needed to drive a sailboat at high speed is to control the overturning and pitching torques caused by the wind, and to get the hull out of the water. If one cared only for speed and had little or no concern for his destination, this would be enough. To have a larger choice in the *direction* of movement at high speed, however, it is essential to consider a sailboat's efficiency.

For immediate purposes, the same definition of efficiency will be applied to the sail as was applied previously to the planing boat and the hydrofoil. We are now familiar with the concept of lift-to-drag ratio (L/D), after having been through a short course in bathtub hydrodynamics. We are not as familiar with the concept of lift when nothing is being lifted, and it is not appropriate to the study of horizontal forces acting on sailboats for other, more serious, reasons. The term, however, will be employed in this chapter only as a convenience in developing an important point.

The principal advantages of a sail are light weight, low cost, and the ease of reversing its curvature. The lightness and cheapness of sails are such exclusive attributes that some aircraft designers seeking these characteristics have turned to the sail for a solution. The sail also solves the problems of collapsibility and storability. Unfortunately, the best L/D we can get with a single sail is about six, and when the aerodynamic drag of the rest of the boat is taken into account, it drops to about four. Only when all the running rigging is passed through hollow, streamlined masts, and the deck is cleared of discontinuities (including people), can the L/D be raised again; and at a heavy cost in the addition of fancy booms and sails. The gains are small compared to the trouble involved, and the whole process is marginal from beginning to end.

Defect in The Sail

The defect in the sail is as inherently incurable as the defect in the hull. The paramount obstacle is the inability of a sail to take a higher pressure on any part of its leeward side than it takes on its windward side. This means that we have to maintain a wind angle (angle of attack) at least 10 degrees higher than is possible with a good airfoil or we will lose the sail's curvature (Figure 40). It also means that we cannot apply the aerodynamic principles that yield high L/D, even at high angles of attack, because these also depend on the existence of rigid, high-pressure surfaces on the leeward side. The net result is simply [1]that there is no forward component to a sail's force until the wind is about 25 degrees off the bow.

The combined L/D of hull and keel (or centerboard) is better, but it is not really a cause for complacency. When tilted by the wind and struck continuously by wind waves, even a good sailboat drifts about ten or more degrees off course. The L/D's of sail, hull, and keel taken together show that at boat speeds that are very low in relation to wind speed (the most favorable condition for sailing close to the wind), a sailboat cannot[2] progress to windward at less than 35 degrees off the true wind. At first glance this prospect may not appear forbidding, but suppose we attempted to make higher speed on this point of sailing. The sails would luff immediately, for the increased speed of the boat would produce, on the sails, an equal wind that would be 35 degrees off the old wind. The boat would have to turn off the wind still more to fill the sails again. The faster it moved, the more it would have to turn off

the wind. If the boat were to move at the speed of the wind, it would find itself in the ridiculous position of sailing at 70 degrees from the true wind in order to keep its sails filled. To get the power needed for such velocities, the angle would certainly have to be raised by at last 10 degrees. Not much progress to windward can be made at such an angle. A sailboat makes much more progress with a third the speed at an angle of 45 degrees off the true wind.

However, a lookout standing on the bow of a speeding sailboat who feels a wind that is perhaps 25 degrees off the boat's axis, can delude himself that the wind has increased in velocity and that the boat is pointing close to the wind. After apparently beating to windward all day, the pilot may come into an anchorage and report with an honest face that he was delayed by mysteriously heavy currents. Now we can understand the confusion of some catamaran sailors, referred to in Chapter VII.

Illusion of Speed

Actually the illusion is even greater than described above. From his place at the tiller, the helmsman feels the wind after it has bounced off the sails, sometimes straight toward the stern. Moreover, the direction in which a boat points is not necessarily the direction in which it moves—it could also be drifting to leeward. Between the direction of the apparent wind and the apparent direction of the boat, the sailor in a fast sailboat might easily be convinced that he is going upwind when he is actually going downwind. In all this, one must not forget that the sailor who wants very badly to move upwind tends to favor the sensory impressions supporting his hopes, while diminishing those that do not. He does not always pay attention to the wind waves, which are the only references he has to the true wind. In confused seas these, too, may become indistinguishable.

To what extent can this situation be improved? Not much can be done to relieve the

---

[1] *These lines should read:* that there is no substantial forward component to a sail's force until the relative wind is about 25 degrees off the boat's direction of motion.

[2] *Insert* make substantial *before* progress

confusion of some sailors, although great improvements can be made in windward sailing if we abandon the sail and centerboard.

Unfortunately we do not have simple bathtub-type experiments at our disposal to aid us in arriving at this solution. It is necessary to borrow from the highly-refined store of knowledge called fluid dynamics. Furthermore, we must resort to our own devices to avoid the traps that await those who attempt to extrapolate too freely from one discipline to another. It turns out that the components of the forces, as hypothesized and carefully resolved in fluid dynamics, have only limited use in analyzing the horizontal forces that act on a sailboat.

Consider what takes place when sailing on a tack: The sails are trimmed in a definite relationship to the centerboard, which provides the load that is opposed by the sail. This relationship can be held regardless of the angle of attack on the sails. As we shall see later, a very fast sailboat would rarely require a change in the orientation of sail to centerboard, the angle of attack notwithstanding. If the wind comes more from the bow, the adjustment can be made simply by turning the whole boat to leeward. If the wind turns more abeam, we can, without tampering with the sail, change the angle of attack by turning the boat to windward.

Note how different this is from the operation of an airplane, which cannot change the direction of the force exerted by its load. The lift can be changed only by changing the angle the wing makes with the induced wind, and this cannot be accomplished without changing the wing's orientation to the load. In this case it is of fundamental importance to measure the force developed to resist gravity—that is, the *lift* created by the wing. Not so for the sail.

Return to Fundamentals

Since this exposition has been somewhat complex, a return to fundamentals may clarify the argument. What we need to know is the direction and magnitude of the sum of all the forces acting on a foil when the flow is at various angles to the foil. The sum of the forces is called the resultant; examining the dimensions of the resultant gives us a simpler, more vivid (and certainly more realistic) picture than does examining the components or the ratio of components only. Figure 41 shows these resultants for an airfoil as applied to both a boat and an airplane. Unfortunately, such diagrams are rarely found in texts on aerodynamics. I am reversing a fifty-year trend by going back to the forces from which the components are derived.

Contrast with Airplane Wing

Figure 41 illustrates the great difference between the behavior of the airplane foil and the foil that serves as a sail. The resultants from an airplane foil move backward, with respect to the direction of the load, as the angle of attack is increased; but the resultants from a boat foil move forward. The benefit to the airplane is greater lift, which is paid for with increased drag. The benefit to the boat is twofold: not only does the magnitude of the resultant increase, which increases the boat's speed, but the changed direction *also* produces higher boat speeds; and there is no price to pay.

In both cases the direction and magnitude of the resultants change uniformly with angle of attack from about two to about 20 degrees. Beyond this the change in direction of the resultant from the airplane foil increases at a greater backward rate, while, at the same time, the magnitude drops. This is of great importance to an airplane because it signals the onset of stalling, with a consequent loss of lift and a great increase in drag, as we can see in the figure. But it has nowhere near the same significance for the boat airfoil. In this case the resultant is brought back to about where it was at 16 degrees angle of attack, also evident from the figure.

As defined in classical aerodynamics, lift is measured in a direction perpendicular to the wind and drag in the same direction as the

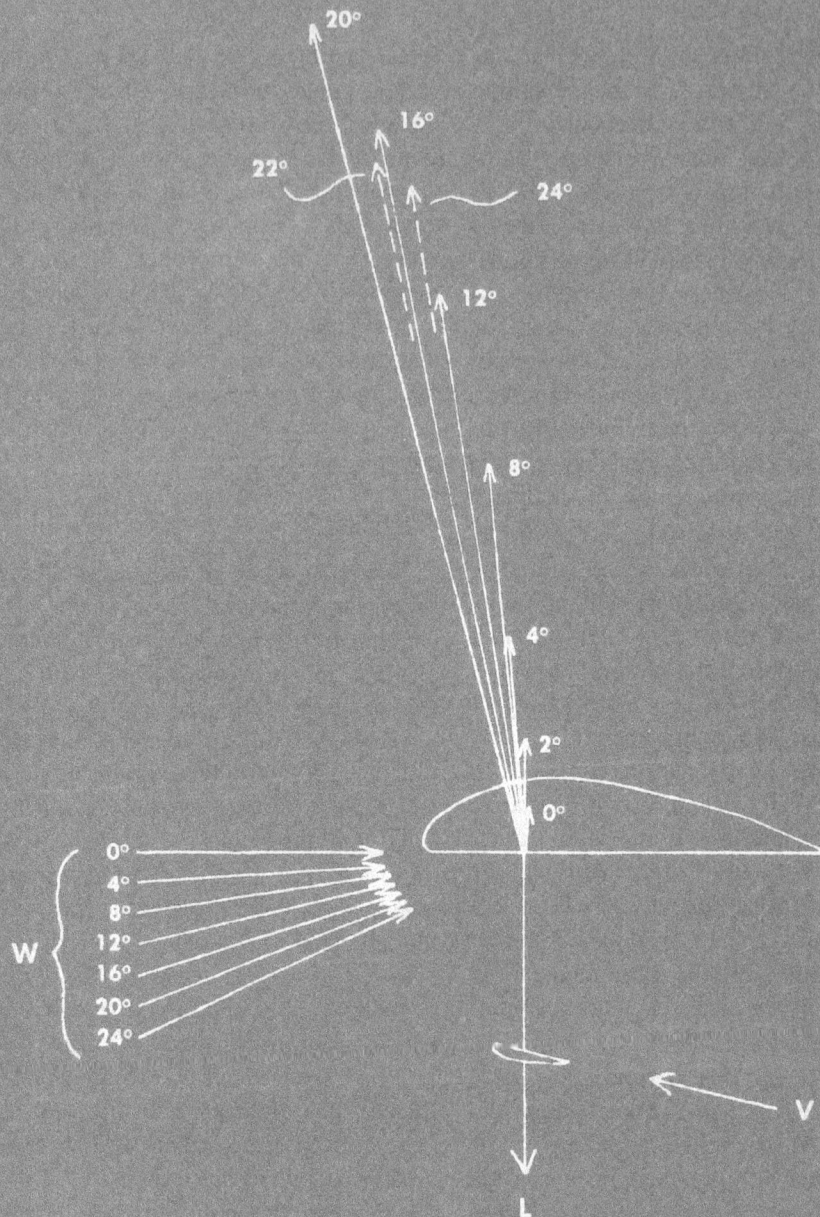

FIGURE 41A

Airfoil Forces

(APPLIED TO SAILBOAT)

This diagram illustrates the fixed relationship of the sailboat's airfoil to the load (L) generated by the keel. Changes in airfoil angle of attack are not necessarily accompanied by changes in the orientation of one with respect to the other, although the direction of the wind (W) can be varied with respect to the direction of motion (V).

For the sailboat the resultants point more in the direction of motion

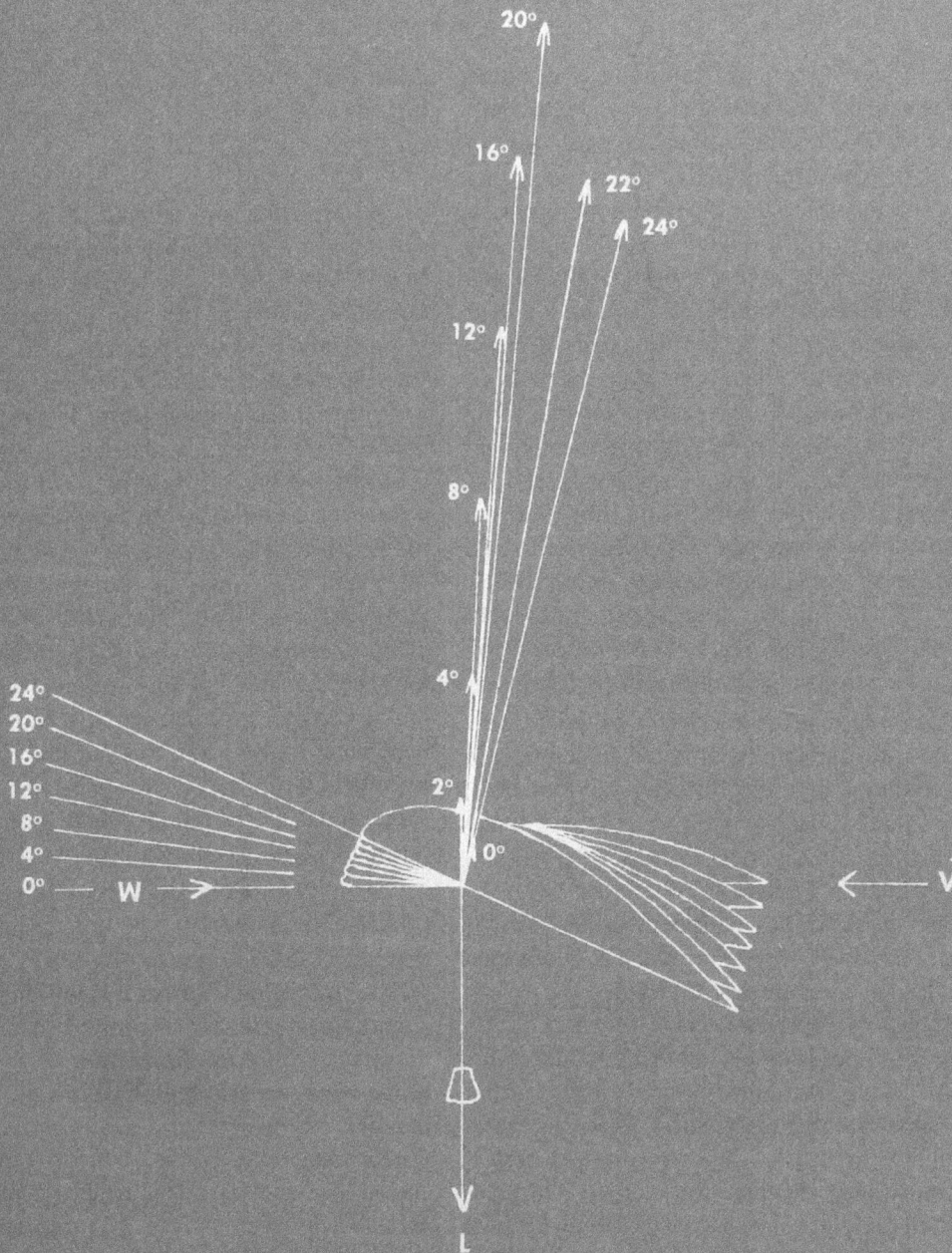

FIGURE 41B

Airfoil Forces

(APPLIED TO AIRPLANE)

The airplane wing, above, cannot change its angle of attack without also changing its orientation to the load, which is the weight of the loaded airplane. The direction of the wind, however, almost always remains fixed with respect to the airplane's motion. The numbered vectors show how the angle of attack in each case is related to the resultant force developed by the airfoil.

as the angle of attack increases. The reverse is true for the airplane.

wind. These directions would have meaning for the sailboat only if it could sail directly into the wind, which is impossible. The most logical line of reference for resolving the aerodynamic forces on the sailboat is the direction of motion in the water. Unfortunately, neither sail nor centerboard hold a fixed orientation to the boat's motion on all points of sailing. Angle of attack changes for both appurtenances depending on whether the boat is running, beating, or reaching. Consequently, each point of sailing must be looked at separately for the interactions of the air and water resultant forces. As the boat sails closer to the wind, however, the aerodynamic components, perpendicular and opposite to the boat's motion, more nearly coincide with lift and drag, and therefore, for sailing close-hauled, the L/D of sail or airfoil is a reasonably valid measure of performance.

The criterion for efficiency in beating to windward is the smallness of the angle between the resultant and the direction of the wind. This angle is smallest at angles of attack giving the highest L/D, about 4 degrees for the foil represented in Figure 41. For other points of sailing the L/D has considerably less significance. Obviously for an airplane it is of equal significance at all angles of attack.

A boat can operate very well with an airfoil that would not be at all suitable to an airplane. A foil giving resultants that move backward at the same rate that the angle of attack increases would be useless for modern aircraft, even if the magnitude increased for another 10 degrees beyond the stall point. Such a foil could be quite appropriate to a boat, provided that the initial resultant at an angle of attack of, say, 4 degrees was as good as the one we have been examining. For the boat the resultants would be piled on each other, holding the same direction, but changing only in magnitude. We would lose some relative speed on a beam reach with this imaginary foil, but any sailor would agree that we could afford to pay this price if it gave us more speed to windward and

downwind, the poorest points of sailing for present boats.

Aspect Ratio

The direction and magnitude of a foil's resultants do not depend entirely on its profile. In addition to other factors, the resultants are critically related to the plan of the wing and, in particular, to the aspect ratio (Figure 42). The aspect ratio is most easily described for a wing with a rectangular form. In this case the aspect ratio is simply the length (span) divided by the breadth (chord). For the foil we have been examining, the resultants are valid only with an aspect ratio of six. If the aspect ratio had been less, the angles between the resultants and the wind would have been greater; correspondingly, if the aspect ratio had been greater, the angles would have been less. In short, the efficiency of a foil is highly dependent on how long we can make the span in proportion to the chord. We saw much the same thing in converting a knife from a planing boat to a hydrofoil. It would be perfectly proper to attribute a large measure of the improvement obtained in the conversion to the increase in aspect ratio.

The aspect ratio of a tapered wing is only slightly more difficult to compute. One merely divides the span by the average of the chords at the wing's widest and narrowest parts. More complicated plan forms naturally have more complicated expressions, but we can bypass these. The big gains in achieving maximum L/D occur when the wing is tapered toward its tip to about a third of the maximum chord. A little more is gained by rounding the tip.

A sail, of course, is only half a wing, and so is a surface-piercing hydrofoil. If circulation from the high-pressure side to the low-pressure side can be prevented around the roots of such half wings, the aspect ratio would be effectively[3] doubled because the flow over the half wing would correspond to the total flow that would take place if the other half were present. Keels,

---

[3] *For* effectively doubled *read* greatly increased

centerboards, and sails that extend right down to the deck automatically achieve the doubled[4] aspect ratio.

The resultants also depend on skin roughness, the viscosity, density, and velocity of the fluid, and the actual length of the chord. There is little or nothing we can do about any of these factors except for skin roughness, which is best reduced by a generous application of elbow grease. The wind's viscosity, density, and velocity will simply have to be what we find them on the day we try the boat, and the airfoil's chord will have to be proportional to the size of the boat. We will not go too far astray by ignoring the influence of these parameters in this analysis of the foil.

With this brief summary, we go on to a synthesis and an analysis of the high-speed sailboat that we have named the aerohydrofoil.

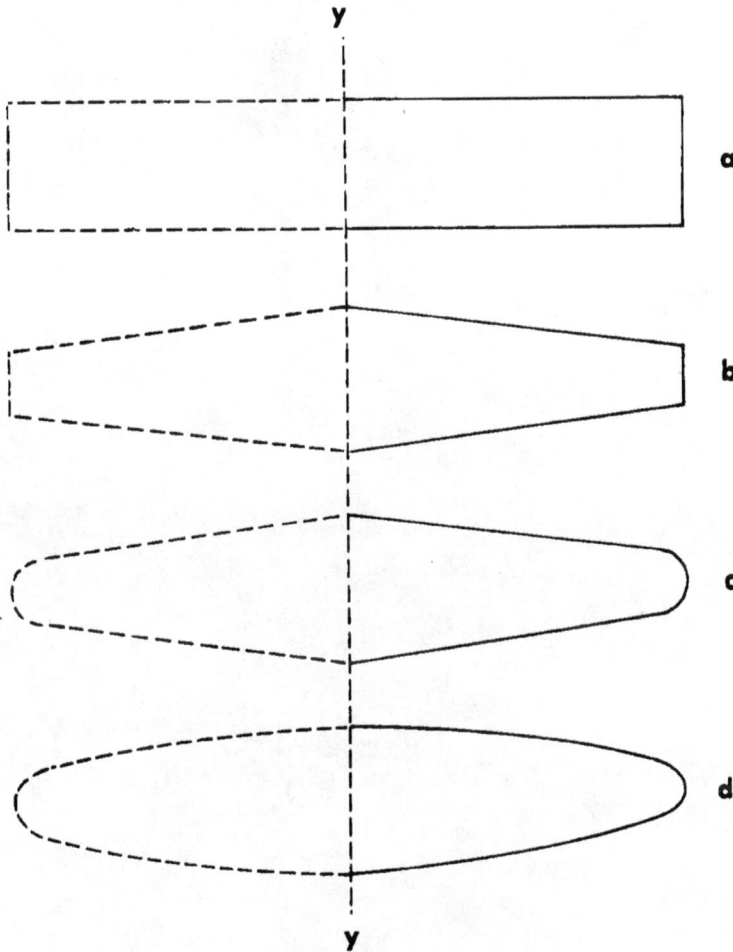

FIGURE 42
Plan Form and Aspect Ratio

The aerodynamic efficiency of a wing is poorest when the plan form is rectangular and greatest when it is elliptical. Plan form (c) comes close to the efficiency of an elliptical wing and is a reasonably uncomplicated design for construction, although not so simple as the rectangular plan form. Sails and hydrofoils frequently have the plan form of half wings, such as those shown with solid lines. If a barrier is placed along the line Y-Y, such that no circulation is possible from one side of the wing to the other, around Y-Y, the aspect ratio of the half wing is equivalent[5] to that of a full wing.

---

[4] For the doubled read an improved

[5] For is equivalent to read may be nearly

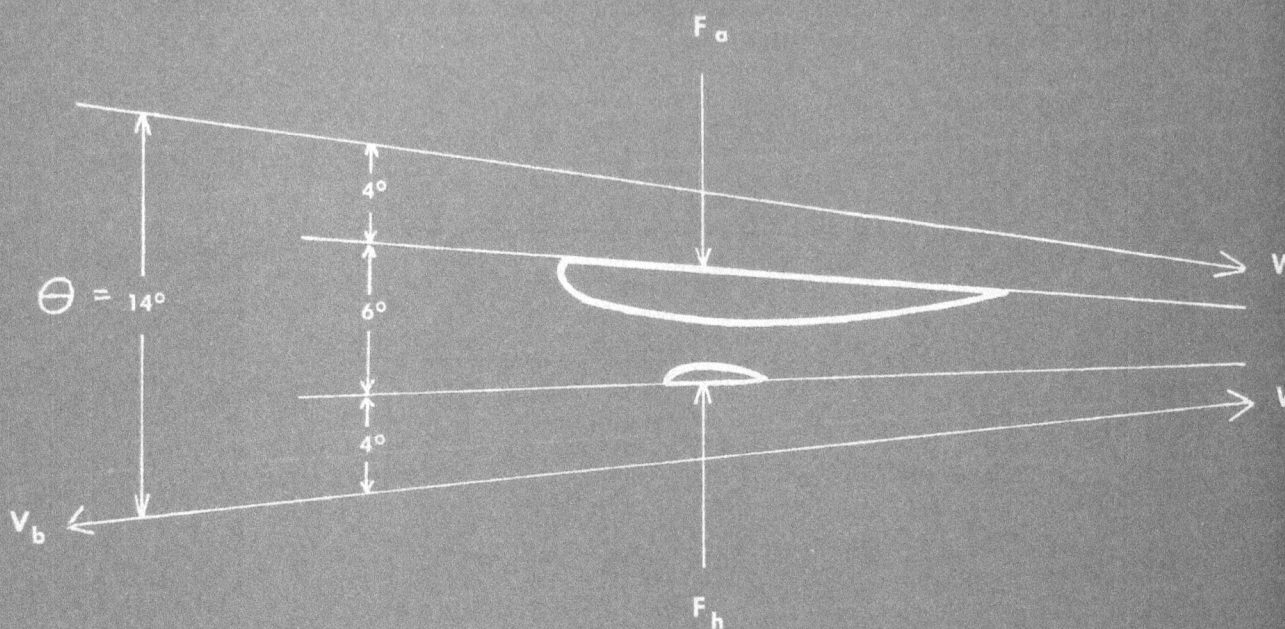

FIGURE 43

Foil Forces

The view shows horizontal sections through an airfoil (F_a) and a hydrofoil (F_h). The foils are rigidly connected, and if they are properly designed, the resultant forces from each can be made to balance when the direction of the boat's motion (V_b) is 14 degrees off the direction of the relative wind (W_r). The L/D ratios of the two foils as discussed in the text are measured from W_r and W_b. Should the ratios be greater than 10, the angle between foil chord lines, shown as 6 degrees in the figure, could be reduced, thereby reducing θ. However, because the analysis is not valid unless flow conditions are established for the hydrofoil, the initial 6-degree angle between chord lines is dictated by the need first to induce motion in the water.

CHAPTER IX

The Theory of the Aerohydrofoil

We have learned thus far that to have a truly high-speed sailboat we must get rid of the boat and the sail. This statement has a slightly absurd ring and probably sounds more like a complaint than a piece of information. It would indeed be absurd if we had no inkling of what we could substitute for these primary components, but now we have some ideas that will help us to identify suitable substitutes. The first step is to create a theoretical model with which to determine the limits of performance. This can be done by marrying a vertical airfoil to a vertical hydrofoil in order to learn what would result if only the force of the wind and the horizontal resistance of the water were involved.

I have taken a few liberties in Figure 43, which shows an airfoil developing a force that is resisted by a hydrofoil. The foils are displaced from each other horizontally merely for convenience in examining the forces and the flow. In all likelihood the hydrofoil would be under the airfoil; but no matter—it is only essential to think of them as being rigidly attached in some specific orientation and to think of the opposing forces as being in equilibrium.

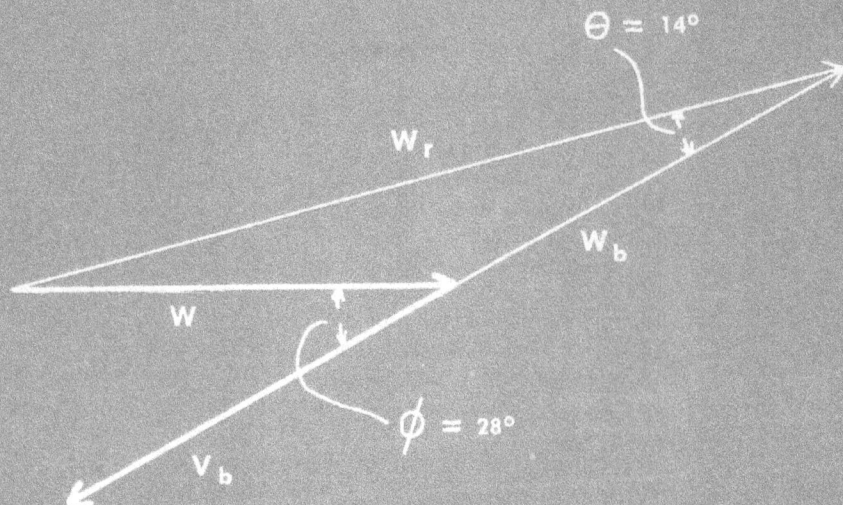

$\Theta = 14°$

W_r

W_b

W

$\phi = 28°$

V_b

FIGURE 44

Velocity Diagram

If the conditions in Figure 43 can be satisfied, and the ratio of the airfoil area to the hydrofoil area is 200 to 1, the boat will be capable of moving upwind as fast as the true wind, at about 28 degrees off the true wind. W_b in the diagram is the reverse wind induced by the motion of the boat. W_r is the vector sum of the true wind (W) and the induced wind (W_b), and it is the apparent wind experienced by an observer on the boat.

FIGURE 45

Aerohydrofoil

Under the conditions described in Figure 44, each of these two foils, one in the air and the other in the water, develops equal and opposite forces. The areas shown in this illustration satisfy the ratio required by the different densities and different velocities of the two fluids.

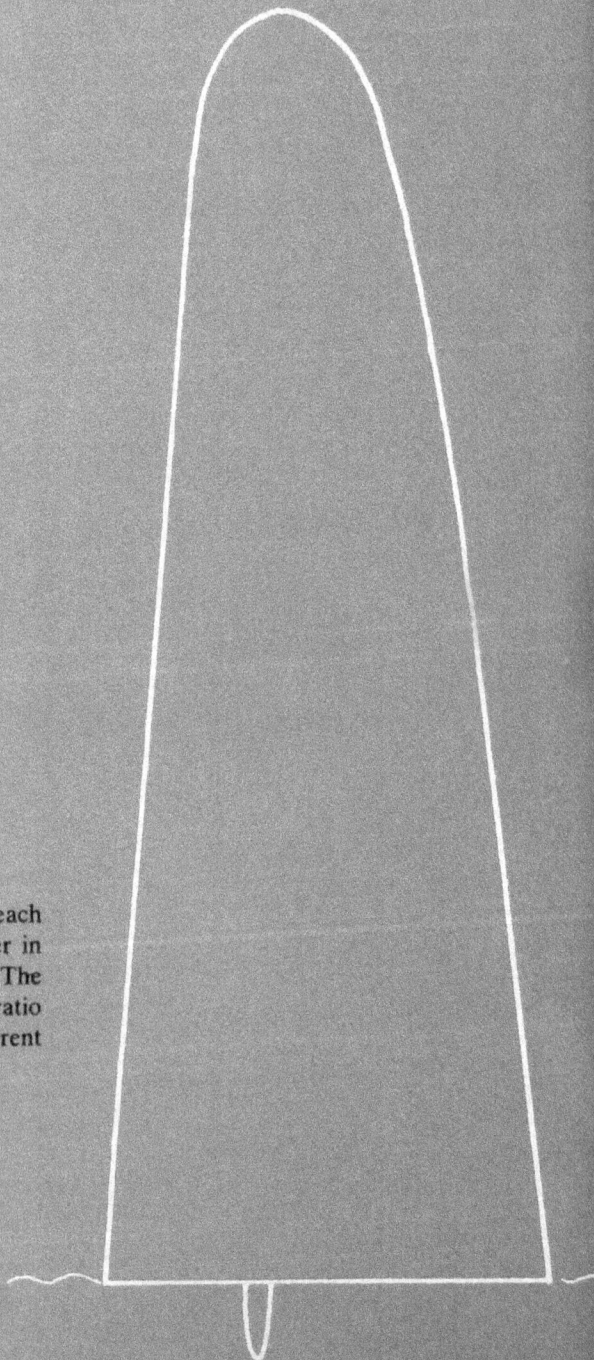

Since our first purpose will be to find the smallest angle possible between wind and way, it is obvious that our quest is for conditions that will yield the smallest angle between the direction of flow and the resultant force from the foils, as well as the smallest angle between the chord lines of the airfoil and the hydrofoil. As stated in Chapter VIII, these conditions are best fulfilled for beating to windward at an angle of attack that gives the highest L/D (measured from the direction of fluid flow). Therefore, with very good foils having L/D's of 20 or greater (usually attained at an angle of attack of about 4 degrees) we should be able to achieve equilibrium of forces with the chord lines almost parallel and the direction of motion in the water a mere 8 degrees off the relative (apparent) wind!

Unfortunately, an aerohydrofoil designed in this fashion knows nothing of the logic behind the design, so it is not at all surprising to find that the foil is not influenced by the design. I have found from bitter experience with scale models that such an aerohydrofoil behaves most uncertainly. Although the logic is correct as far as it goes, it does not take into account the conditions prior to the development of proper flow around the hydrofoil. The resultant developed by the hydrofoil when moving broadside, or, for that matter, when moving backward at a 4-degree angle of attack, is in essentially the same direction as the resultant developed when moving forward at a 4-degree angle of attack. The boat simply does not know which way to go until the angle of attack on the airfoil is very high; a situation entirely unsuitable for high-speed beating to windward. In essence the hydrofoil is a vertical glider wing, and therefore it must have a glide angle before it can move forward and develop the needed flow.

To ensure that flow will start as soon as the first useful resultant is derived from the airfoil, the chord line of the hydrofoil has been placed at slightly more than a right angle to the airfoil's resultant at a 4-degree angle of attack. This also helps the hydrofoil to glide at the angle of attack for maximum L/D, which is desirable for beating to windward. Taking all this into account, plus a little more drag than theory grants, we have a total angle of 14 degrees between relative wind and way, provided the total L/D for the foils and other boat surfaces is not less than about 10 at a 4-degree angle of attack. This assumption takes no advantage of the better conditions that are possible with foils of higher L/D after the boat is under way. However, it is a conservative concession to practicality, and it also serves greatly to simplify the analysis that follows.

Balance of Forces

The vector diagram in Figure 44 shows that such a boat would be capable of moving 28 degrees off the true wind as fast as the wind. This, of course, is most encouraging, but we have not yet considered how to account for the difference between the velocity of the apparent wind and the velocity of the water flow, or the difference between the density of the air and the density of the water. Until we have taken care of the forces on the foils, so that they are truly in balance under the conditions appropriate to each foil, the diagram is not valid.

This balance can be achieved in a fairly simple way. The apparent wind (W_r) in Figure 44 has almost twice the velocity of the water flow. We recall that inertial drag is proportional to the square of the velocity. A foil's resultant follows the same law. Hence, if both foils were in the same medium, one moving twice as fast would create four times the resultant force of the other. It is necessary only to reduce its area to a fourth that of the slower foil in order to make the forces equal. But one foil moves in water and one in air, which at sea level is only 1/800 the density of water. The great equalizer here, too, is area. We cold-bloodedly increase the area of the airfoil 800 times, after having cut it to one quarter, making its final area about 200 times that of the hydrofoil.

The 40-Knot Sailboat

Idealized Aerohydrofoil

One more step, and we can picture the idealized aerohydrofoil as it appears from the side. If both foils have the same plan form and one has 200 times the area of the other, the ratio of the spans is the square root of 200, or, to a sufficient approximation, the airfoil has 14 times the span of the hydrofoil. Figure 45 illustrates the result. It is obviously preposterous—the displacement of the hydrofoil could never support such a structure. The assembly is unstable about every axis. The mechanisms for reversing the curvature in both foils, essential to changing tacks, would be a nightmare of engineering. Nonetheless, this idealized aerohydrofoil provides an excellent basis for establishing the performance limits of sailboats that are subject only to horizontal forces. By applying some mathematics to its properties, the velocities achievable on all possible points of sailing can be derived.

The index of flow conditions, or the Reynolds number (R), will be disposed of next. This quantity is defined as:

$$R = \frac{Vl}{v}$$

in which V is the fluid velocity, l is the characteristic length of the body in the flow path, and v is the kinematic viscosity. Should the Reynolds numbers of the two foils differ greatly, the application of the same coefficients to both foils would be questionable. In the absence of specific values for V and l, we may employ the *ratio* of Reynolds numbers as a basis for judgment, providing the true Reynolds values are over 10,000, which is assured. This ratio can be derived from the information developed thus far.

Under the initial conditions described for the aerohydrofoil,

$$\frac{W_r}{V_h} \cong 2$$

The ratio of the characteristic lengths is effectively the ratio of the spans; thus

$$\frac{l_a}{l_h} = 14$$

The kinematic viscosities of air and water at sea level, and at 70° Fahrenheit, are

$$v_a = 1.64 \times 10^{-4}$$
$$v_h = 1.02 \times 10^{-5}$$

giving a ratio of 16. When combined, these values yield a Reynolds ratio of 1.75. For all other points of sailing the ratios are smaller. We conclude, therefore, that the foils are in the same flow regime and that the dynamic coefficients apply equally to both.

Assumptions

The remaining assumptions used in the calculations are listed below:

1. The hydrofoil does not ventilate. (This is reasonable for the low angles of attack contemplated and is important primarily for beating to windward at speeds that will almost always be subcavitating.)

2. Splash losses on the hydrofoil's leading edge are trivial. (This is not so reasonable for the foil shown in Figure 43, although good hydrofoils do exist with trivial splash losses.)

3. Speeds are always high enough to pass well beyond maximum wave drag. (This is reasonable, considering that a one-foot root chord for the hydrofoil goes with a 14-foot root chord for the airfoil. One foot of waterline gives maximum wave drag at about two knots.)

4. End losses around the roots of the foils are trivial. (This is reasonable if we assume the root of the airfoil is so close to the water that no significant airflow occurs under it.)

5. The driving force developed by the airfoil increases in direct proportion to the angle of attack. (This is almost true until the stall point is reached. By assuming that the airfoil can be turned with respect to the hydrofoil before the stall point is reached, the assumption can be kept true well beyond it. By doing this, more of the airfoil resultant is used to drive the boat.)

6. The airfoil's resultant increases in proportion to the square of the wind velocity.

7. The hydrofoil's angle of attack remains sensibly constant at all speeds. (This is prac-

tically assured because the airfoil's resultant moves forward as it increases in magnitude.)

8. The hydrofoil's resultant increases to match that of the airfoil through increased velocity. (There is nothing else it can do at a constant angle of attack. This is important primarily for beating to windward; otherwise it does not matter much.)

9. The hydrofoil's resultant is proportional to the square of the water velocity.

Velocity Vectors

The velocity vectors corresponding to Figure 44 are:

> True wind = W
> Wind induced by the boat = W_b (equal and opposite to V_b)
> Wind relative to boat = W_r (vector sum of W and W_b)
> Velocity of boat = V_b = $-W_b$

According to the assumptions, the force exerted by the airfoil (F_a) is proportional to W_r^2 and to the angle of attack ($\propto$), or

$$F_a \sim W_r^2 \propto$$

We can express this relationship in terms of the inital force (F_0), the ratio of W_r to W, and the ratio of $\propto$ to the initial angle $\propto_0$. Thus

$$F_a = F_0 \left(\frac{W_r}{aW}\right)^2 \frac{\propto}{\propto_0} \qquad (1)$$

in which a is a constant factor denoting the ratio of W_r to W under the initial conditions.

But, according to the assumptions, F_a is also proportional to V_b^2. That is, the forces on both foils are in equilibrium, hence

$$F_a = F_0 \left(\frac{-V_b}{W}\right)^2 = F_0 \left(\frac{W_b}{W}\right)^2 \qquad (2)$$

in which $W_b = -V_b = W$ under the initial conditions.

Combining Equations (1) and (2) gives

$$W_b^2 = \left(\frac{W_r}{a}\right)^2 \frac{\propto}{\propto_0} \qquad (3)$$

By trigonometry, we see from Figure 44 that

$$W^2 = W_r^2 + W_b^2 - 2 W_r W_b \cos \theta \qquad (4)$$

Combining with Equation (3), we get

$$\left(\frac{W_b}{W}\right)^2 = \frac{\propto}{a^2 \propto_0 + \propto - 2 a \cos \theta \sqrt{\propto \propto_0}} \qquad (5)$$

Under the initial conditions of Figure 43,

$$a = 1.94$$
$$\propto_0 = 4°$$
$$\propto = \theta° - 10°$$

Substituting in Equation (5) and taking the square root,

$$\frac{W_b}{W} = \frac{\sqrt{\theta - 10}}{\sqrt{\theta + 5.04 - 7.76 \cos \theta \sqrt{\theta - 10}}} \qquad (6)$$

Since W_b is numerically equal to V_b, we now have an expression for the ratio of the boat velocity to the true wind velocity in a form that contains only one variable. When we take into account the directional changes in the airfoil's resultant arising from increases in angle of attack and rotation of the airfoil after the stall point, we see that this expression is sensibly valid for values of θ up to 90 degrees.

W_b/W and θ are entered into equation (3) in order to compute W_r/W. The true angle off the wind (ϕ) can be derived from

$$\frac{W_r}{W} = \frac{\sin (180 - \phi)}{\sin \theta} \qquad (7)$$

The table below presents some representative figures computed to slide-rule accuracy.

$\phi°$	Ratio of V_b/W	$\propto°$	$\theta°$	Ratio of W_r/W
28	1.00	4.0	14.0	1.94
45	1.62	6.7	16.7	2.43
60	2.05	8.8	18.8	2.68
75	2.31	10.6	20.6	2.75
90	2.37	12.9	22.9	2.57
105	2.29	15.3	25.3	2.27
120	2.00	20.0	30.0	1.73
135	1.35	37.5*	47.5	.94
150	1.05	58.0*	68.0	.54

* Theoretical. Real angles of attack would be lower because the airfoil would be turned to accommodate the direction of wind.

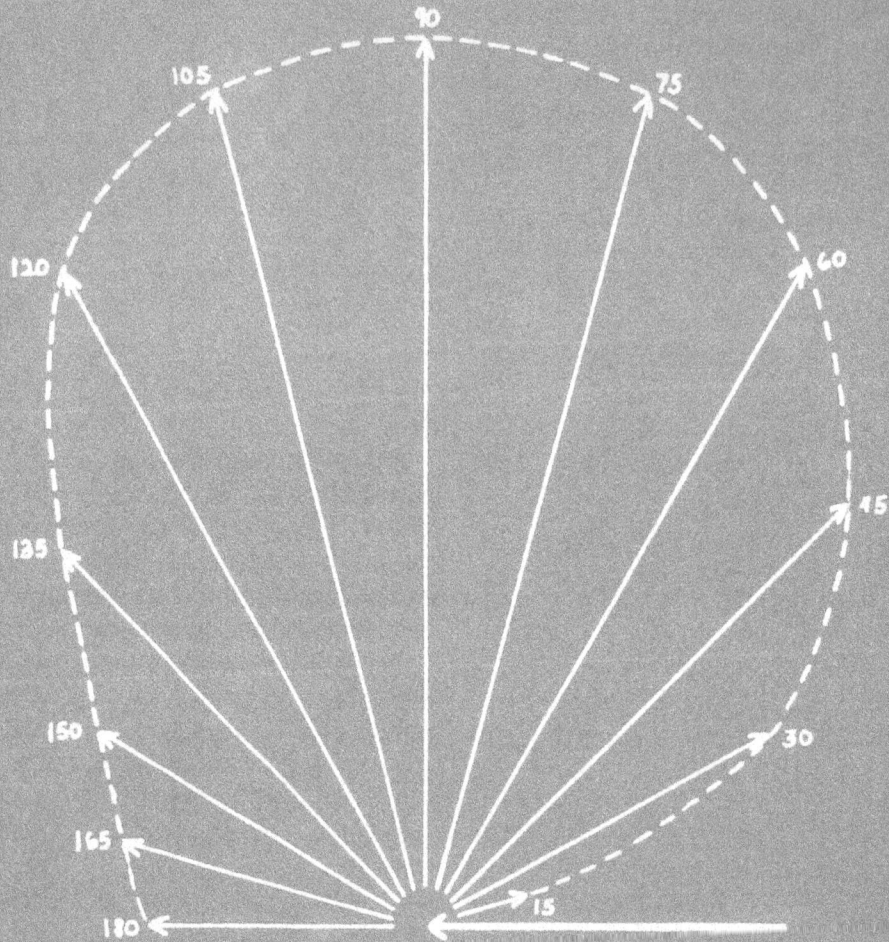

FIGURE 46

Velocity Versus Angle off the Wind

The heavy arrow represents the magnitude and direction of the true wind. The length of the lighter arrows shows the relative velocity of the aerohydrofoil sailing at various angles to the true wind. The 30° vector (upwind) is longer than the 150° vector (downwind). The upwind component of the 45° vector is 10 per cent greater than the wind itself. Port tacks would be represented by a mirror image of the diagram.

Faster than The Wind

A graphic presentation of speeds on all points of sailing is given in Figure 46. Apparently the speed at angles between 30 and 150 degrees off the wind is faster than the wind itself. Even more surprising, the aerohydrofoil goes over twice as fast as the wind between 60 and 120 degrees. This is similar to the achievements of iceboats.

At about 150 degrees the relative wind is broadside to the boat and about half the strength of the true wind. Almost all of the force from the adjusted airfoil is then turned forward. But the speed is no greater than at 30 degrees heading. At 180 degrees the wind is directly from behind, and the driving force is essentially the flat plate drag of the airfoil. In this direction the boat moves no faster than when pointing off at 20 degrees. Therefore, going dead before the wind is generally slower than going upwind. Unless one were in narrow waters, better progress could be made downwind by tacking at 120 degrees. Note from the table that 125 degrees is about the first point of sailing that requires an angular adjustment of the airfoil with respect to the hydrofoil. With a proper slot in the airfoil, even this would be unnecessary.

From Figure 46 it is obvious that the best progress upwind is made by sailing at 45 degrees. However, in close waters the ability to make headway at 15 degrees, even slowly, is important. Of considerable interest is the fact that the average progress upwind, at the best off-wind angles, is more than 10 per cent better than the wind itself.

Our findings can be summarized with this statement: About half the time, the best way of getting where you want to go in a fast sailboat is to move in some other direction.

This is about as far as simple theory can take us. In fact, the theory had to be macerated a little to bring us this far. We know that it has not provided us with a practical boat, only with a delineation of potentialities and limits under certain conditions. It is appropriate now to say something about the limits of the model itself. Actually, we have only analyzed the performance of a model improvised expressly for beating to windward. In doing so we have needlessly burdened the other points of sailing, which do not necessarily require the characteristic of minimum leeway. Incredible as it may seem, the little hydrofoil jutting into the water, doing yeoman's work upwind, is almost an anchor on downwind sailing. To get the last ounce of excellence, it must be lifted as high out of the water as possible when not needed.

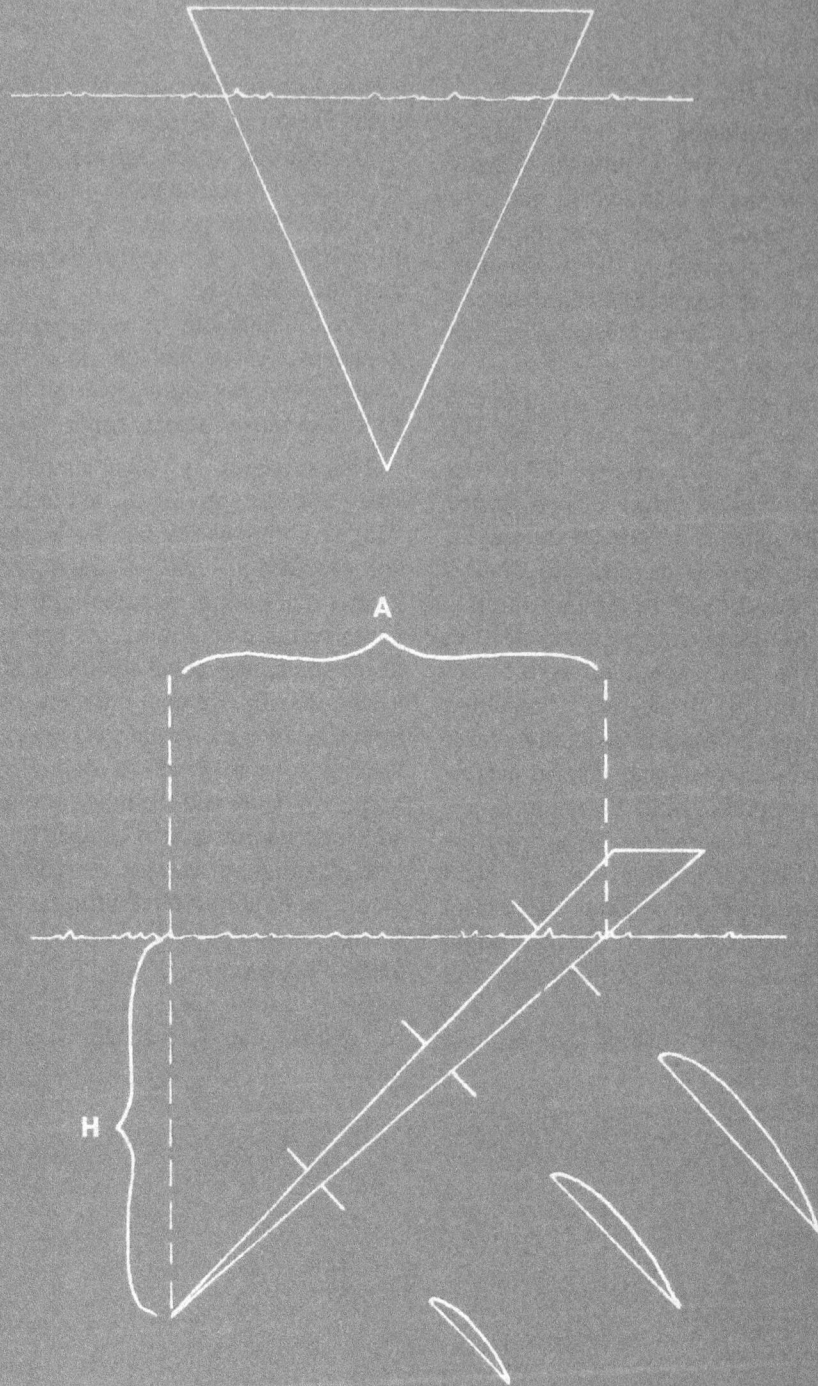

FIGURE 47

Triangular Lifting Foils

Because the sections of a triangular foil are similar from root to tip, and the aspect ratio does not change with depth of immersion, the lift-to-drag ratio remains sensibly constant regardless of depth. In this diagram A is the projected immersed area of the foil and H is the initial depth of the foil.

CHAPTER X

Lifting Hydrofoils

SYMBOLS[6]

A = projected immersed area of foil

A_R = aspect ratio of immersed foil

D = constant of drag proportionality

F_b = buoyant force

F_D = drag force

F_L = lift force

H = initial depth of immersed foil

K_1 = collected constants for foil geometry and water density

K_2 = collected constants for foil geometry and dynamic coefficients

$K_3 = \sqrt{\dfrac{K_1 H}{K_2}}$

$K_4 = \dfrac{D K_1 H^3}{L}$

$K_5 = \dfrac{D K_1 K_2 H^3}{L}$

L = constant of lift proportionality

M = vertical load on foil

R = ratio of immersed depth to initial depth

V = flow velocity

The questions with which we are concerned here are: What happens to the drag of the aerohydrofoil when true lift surfaces are introduced into the design? Is total water drag still proportional to the square of the velocity? To examine these questions a special boat must be hypothesized—a boat free of all complicating horizontal forces, except a driving force in the direction of motion and a drag force opposing the motion. Specifically, the interplay between these forces and the vertical forces of lift and weight will be examined. Here the terms "lift," "drag," and "L/D" can be used in accordance with their precise classical definitions without ambiguity.

The model in this case will rest on special hydrofoils that have a constant aspect ratio, constant angle of attack, and constant thickness-to-chord ratio. Taken together, these properties give constant L/D regardless of depth of immersion. Triangular, surface-piercing hydrofoils with sharp tips would have such properties (Figure 47). The hydrofoil angle of inclination is not pertinent to the analysis

---

[6] *Add to list:*

K = foil geometry factor

For $K_5 = \dfrac{DK_1K_2H^3}{L}$ *read* $K_5 = \dfrac{DKK_1H^3}{L}$

so long as it is neither vertical nor horizontal, but suitably inclined for good flow without ventilation.

Assuming that no lift is supplied by displacement, the lift force is designated as follows:

$$F_L = LAV^2$$

108

But if weight and buoyancy do not change, F_L is also equal to a fixed load (M). Hence,

$$A = \frac{M}{LV^2} \qquad (8)$$

The drag force can be expressed as

$$F_D = DAV^2 \qquad (9)$$

Substituting the equivalent of A from Equation (8) gives

$$F_D = \frac{MD}{L} \qquad (10)$$

This expression does not contain velocity as a factor. It tells us that drag is independent of velocity and is essentially the same at all velocities, provided there is no change in the load or in the ratio D/L.

Power Not Significant

The temptation now is to say that equation (10) is not realistic because it does not take into account the power necessary to drive the boat. If this were to be done, the velocity would come back into the picture, for power equals the quantity $F_D V$. This would be entirely appropriate for a motorboat, which derives its thrust by *reacting* with the water. But for a sailboat the concept of power is no more appropriate than it is for a glider. For either a sailboat or a glider it is necessary to consider only the forces that are in equilibrium in order to determine every factor of importance.

Equation (10) is not realistic for quite different reasons. Several important factors were ignored in the first simple derivation. The one having the largest bearing on the problem is the distribution of the load between displacement and dynamic lift as depth of immersion changes. We shall assume that initially all the lift comes from the buoyant force of the foils, which, while having the same geometry as in the first case, are large enough to support the load statically. The buoyant force (F_b) expressed as a function of depth of immersion, is then

$$F_b = K_1 R^3 H^3$$

in which K_1 is a quantity covering the density of water and the geometry of the foils, R is the ratio of the immersed depth to the initial depth, and H is the initial depth.

The lift that must be supplied dynamically is then equal to the initial displacement less the remaining displacement, or

$$F_L = K_1(1 - R^3)H^3 \qquad (11)$$

The dynamic lift must also be equal to F_L, thus

$$F_L = K_2 R^2 H^2 V^2 \qquad (12)$$

in which K_2 includes the foil coefficients, geometry coefficients, and fluid properties.

Combining Equations (11) and (12) gives

$$V^2 = \frac{K_1}{K_2} \frac{(1 - R^3)}{R^2} H \qquad (13)$$

or

$$V = K_3 \sqrt{\frac{1 - R^3}{R}}$$

in which

$$K_3 = \sqrt{\frac{K_1 H}{K_2}}$$

Since drag is the product of the lift and the reciprocal of L/D,

$$F_D = \frac{D}{L} K_2 R^2 H^2 V^2 \qquad (14)$$

Substituting Equation (13) for V^2 in Equation (14)

$$F_D = \frac{D}{L} K_1 H^3 (1 - R^3)$$

or

$$F_D = K_4(1 - R^3) \qquad (15)$$

where

$$K_4 = \frac{DK_1 H^3}{L}$$

Two-Phase Relationship of Velocity to Drag

We now have V_b and F_D in terms of a common variable. Computing values for V/K_3 and F_D/K_4 gives the following table:

R	V/K_3	F_D/K_4
1.0	0.0	0.0
.9	.577	.270
.8	.874	.488
.7	1.165	.665
.6	1.477	.785
.5	1.870	.875
.4	2.414	.936
.3	3.285	.973
.2	4.980	.992
.1	9.998	.999

These figures are most interesting. They show a two-phase relationship of velocity to drag.

For R greater than .5, drag is essentially proportional to the first power of V. Note that at R = .5, when the foils are half out of the water, the dynamic lift of the immersed part is supporting seven eighths of the load.

Between .5 and .1 the drag is almost independent of velocity. Here Equation (15) says almost the same thing as Equation (10), which was not far off after all. But the profit is really not quite so high in this part of the speed regime when we take into account the properties of a real foil.

If the foil tips are not sharp, which is indeed the most likely prospect, D/L will not remain constant, because the aspect ratio (A_R) will change with emergence of the foil. The induced drag is an inverse function of A_R. Therefore, D/L should be modified by the factor $1/A_R$. Since there is special interest in the foils' characteristics at small values of immersion, at which the effect of $1/A_R$ is very great, we will not be far wrong in lumping the induced and the profile drag coefficients as a common factor of $1/A_R$.

Proportions of Foils

If the tip is cut off the foil so that the tip chord is one-fifth the chord at the initial water line, the original height can be retained for Equation (11), since F_L refers only to the exposed volume of the foil. The term R^2H^2 in Equation (12) must be modified, however. The reference area for the modified foil, as a function of R, is now proportional to $(R^2 - 1/25)$, making Equation (12) read

$$F_L = K_2(R^2 - 1/25)H^2V^2$$

Combining with Equation (11), we get

$$V^2 = \frac{K_1(1 - R^3)H}{K_2(R^2 - 1/25)} \quad (16)$$

$$V = K_3 \sqrt{\frac{1 - R^3}{R^2 - 1/25}}$$

The new reference area and the factor $1/A_R$ must also be applied to Equation (14). Thus

$$F_D = \frac{D}{L} K_2(R^2 - 1/25)H^2V^2(1/A_R) \quad (17)$$

Since A_R equals the square of the span divided by the area,

$$1/A_R = \frac{K_2(R^2 - 1/25)H^2}{(R - 1/5)^2H^2} \quad (18)[7]$$

Combining Equations (16), (17), and (18), we get

$$F_D = K_5 \frac{(1 - R^3)(R + 1/5)}{(R - 1/5)} \quad (19)$$

in which K_5 includes all the constants.

Computing and tabulating as before:

Reference R	True R	V/K_3	F_D/K_5
1.0	1.000	0.0	0.0
.9	.875	.592	.424
.8	.750	.910	.816
.7	.625	1.215	1.195
.6	.500	1.565	1.573
.5	.375	2.340	2.380
.4	.250	2.780	2.810
.3	.125	4.410	4.865

[7] For $1/A_R = \frac{K_2(R^2-1/25)H^2}{(R-1/5)^2H^2}$ read $1 A_R = \frac{K(R^2-1/25)}{(R-1/5)^2}$

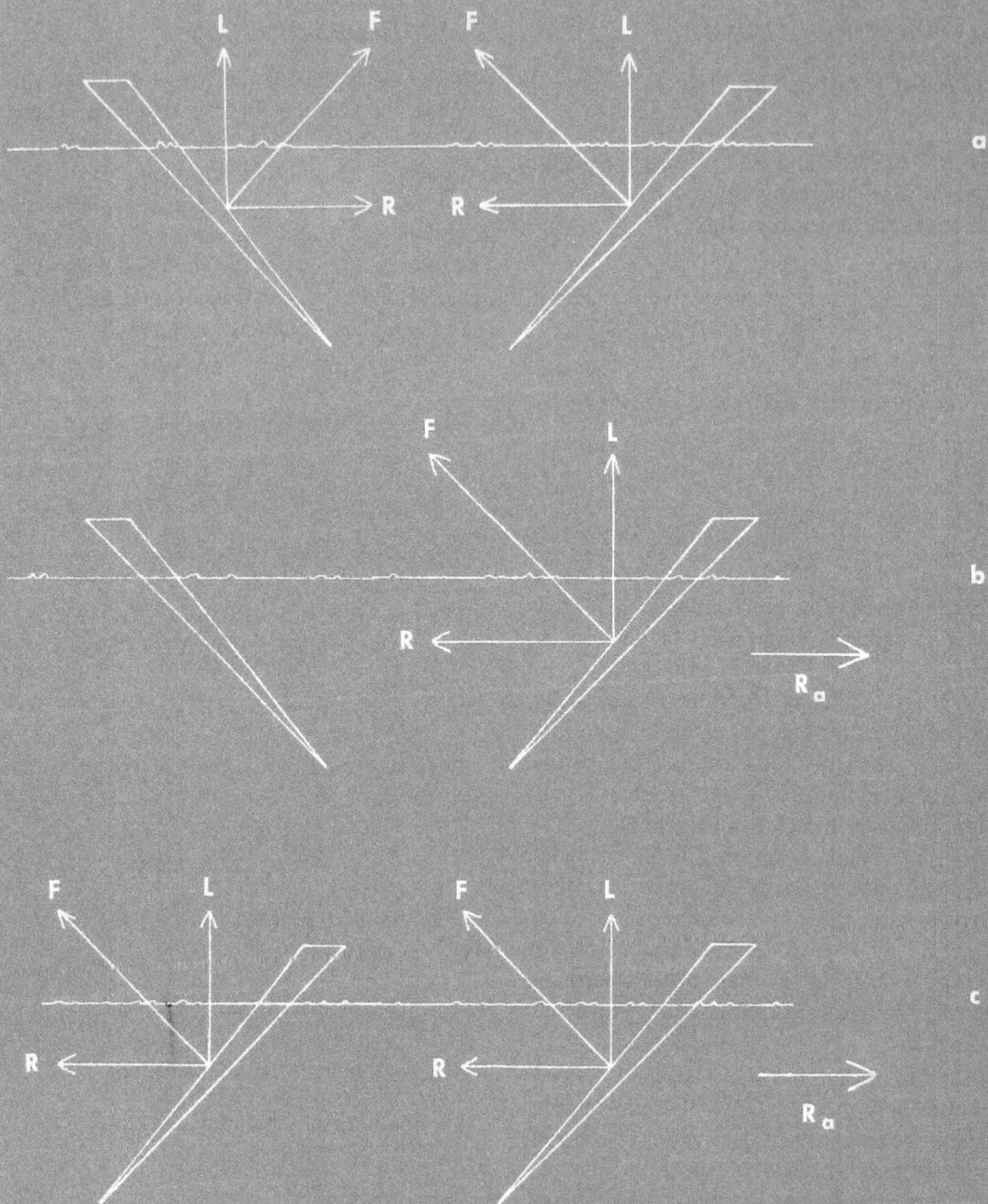

FIGURE 48

Symmetry and Dyssymmetry of Inclination

No side forces act in diagram (a). The horizontal components (R) from each foil cancel each other, leaving only the lift components. This diagram would be applicable to motor driven hydrofoil boats. In (b) the side force (R_a) developed by the addition of a sail as the driving force produces a higher angle of attack on the leeward hydrofoil and diminishes the angle of attack of the windward hydrofoil. For the condition shown, the windward foil has zero angle of attack and can produce no useful force. It yields only a drag force. When the foils are all inclined in the same direction as in (c), they are used most efficiently.

On this basis, drag is almost directly proportional to velocity. (One must bear in mind that the quantities in the second table do not correspond precisely to those in the first table because the foils have different plan forms and because the constants have not been combined in the same way.)

We conclude, by comparing the tables, that where expected immersions are considerably less than 50 per cent, the foils should be fairly sharp; otherwise the more conventional plan forms with better initial L/D are to be preferred. As a matter of fact, the better L/D is needed more on a beat to windward, where the speed is lower and the immersion greater. The added drag introduced by leeway resistance is greatest on this point of sailing. Farther off the wind it becomes less significant, producing less load on the foils and consequently less drag.

Thus for a foil of reasonable plan form, with the tip chord not more than about one-fifth the root chord, the assumption that total drag increases with velocity is fairly valid for most of the velocity range. Further generalization is not warranted, owing to the critical influence that the particular foil may exert on all the factors discussed.

Asymmetrical Arrangement

The subject of hydrofoils for sailboats is by no means exhausted with this quantitative analysis. Other factors are also critical. For example, a symmetrical arrangement of foils, as applied to motorboats, has limited use for sailboats. The reason is elusive, but simple.

When hydrofoils are inclined symmetrically about the center line of a sailboat, as shown in Figure 48b, the wind creates side forces that increase the lift from the leeward foils and decrease the lift from the windward foils. This occurs because the angle of attack must increase on one side and decrease on the other. This undesirable situation can be avoided with only one possible symmetrical arrangement: by holding the hydrofoils in an absolutely horizontal orientation. But this orientation would provide no effective resistance to leeway.

The usual "V" arrangements of motorboat hydrofoils would work well for sailboats going with the wind or on a broad reach. All the foils would then supply lift in the right direction, and the side components would be mutually cancelled. But unless some additional vertical surface were provided, such foils could generate little resistance to leeway until either the angle of attack or the immersion of the leeward foils became much greater than that of the windward foils.

A rather heavy price would have to be paid to get resistance to drift in this way. Depending on the magnitude of the side load, the windward foils could perform in only one of three possible modes: (1) give lift and increase the leeway, (2) give neither lift nor resistance to leeway, only drag, (3) resist leeway and also develop negative lift. Each mode has serious disadvantages. They could be avoided only if the windward foils were out of the water, a condition the boat would reach just before capsizing.

The solution is to have all the foils inclined in the same direction. Aligned with their upper ends to leeward, all the foils would simultaneously supply lift and resistance to leeway, regardless of the wind's force. This is not an ideal arrangement for sailing directly with the wind, but the sailor of a fast hydrofoil boat would never sail directly downwind if there were any other choice. As we have seen, it can make its best speed downwind by tacking.

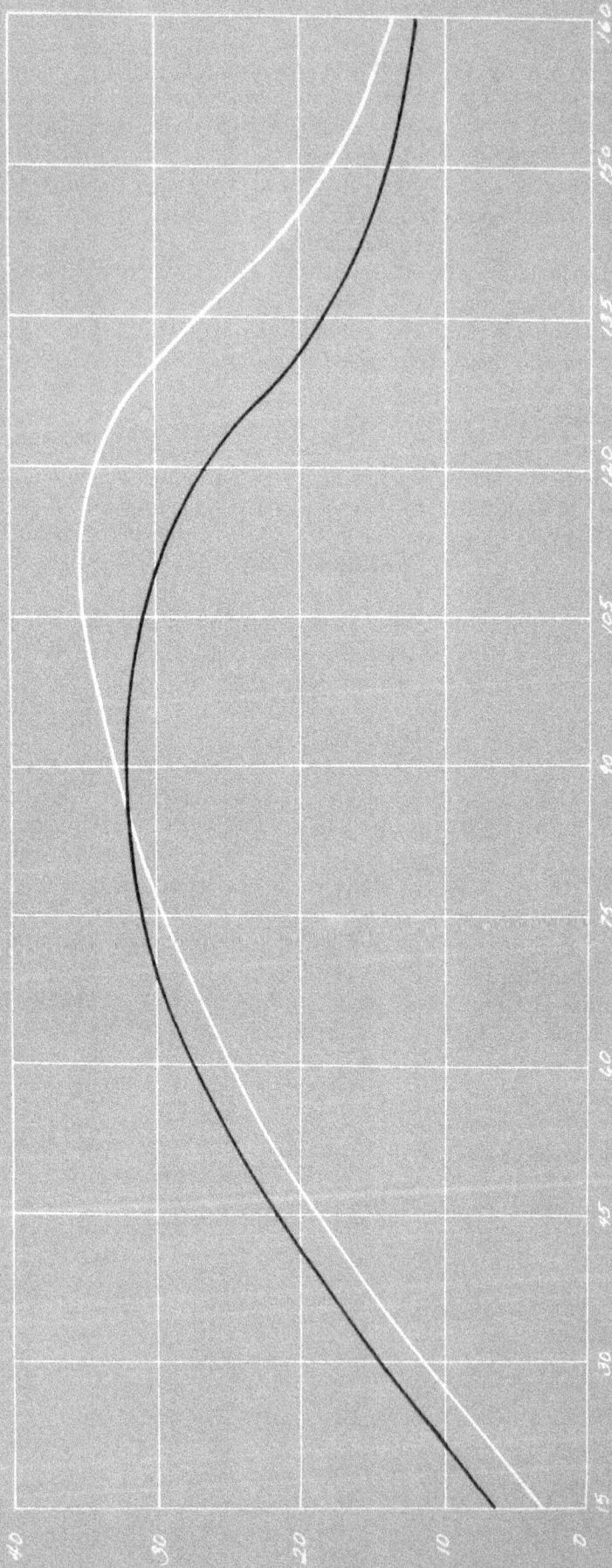

FIGURE 49[8]

Comparison of Lifting and Non-Lifting Aerohydrofoils

The curves show the velocities achieved for each kind of aerohydrofoil in a wind of 13.4 knots. The non-lifting version is moderately better sailing upwind. For off-wind angles greater than 80 degrees the lifting version is superior, and it is decidedly superior at 130 degrees (almost 10 knots faster).

[8] Abscissa is "Offwind Angle," last value is 165 degrees. Ordinate is "Velocity" (Knots). Black curve for non-lifting aerohydrofoil; white for lifting version.

CHAPTER XI

The Lifting Aerohydrofoil

SYMBOLS

SYMBOLS

A = foil area
C_L = hydrofoil lift coefficient
F_L = hydrofoil lift force
M = vertical load on foil
Q = dynamic constant for water
V_b = boat velocity
W = true wind velocity
W_r = wind relative to boat
α = foil angle of attack
θ = angle between boat direction and apparent wind
ϕ = angle between boat direction and true wind

Subscript (a) refers to airfoil.
Subscript (h) refers to hydrofoil.

The influence of dynamic lift on the performance of the aerohydrofoil remains to be appraised. As before, we will first establish the minimum angle at which the boat can sail upwind as fast as the wind. To simplify the analysis, all the hydrofoils will be considered as inclined in the same way, at 45 degrees. Accordingly, the lift and leeway-resisting forces will be perpendicular and equal.

Since the drag is proportional to the true area of the foils and the anti-drift force is proportional to the projected area, the initial drag is about 40 per cent greater than for the non-lifting aerohydrofoil. To retain equality of wind and boat speeds, therefore, the force developed by the airfoil must be increased *in the forward direction* by about 40 per cent. This can be done by increasing the included angle between airfoils and hydrofoils from the original 6 degrees to 8.5 degrees. However, a better arrangement would be to increase the included angle by two degrees and the angle of attack on the airfoil by one degree. This gives a suitable increase in the airfoil's resultant, to compensate for the reduction in W_r attending the increase in θ.

Equality of wind and boat speeds can then be attained if

$$\alpha_a = 5°$$
$$\theta = 17°$$
$$\phi = 34°$$
$$W_r = 1.91\ W$$

To determine precisely *what the speed is* when equality of wind and boat speeds is reached, we must compute the boat velocity at which the hydrofoil area satisfies the ratio established in Chapter IX. In other words, there exists a lift force at a particular V_b that matches the horizontal hydrodynamic force required by the aerohydrofoil. This force corresponds to a ratio of 200 in area of airfoil to area of hydrofoil.

Foil Lift Equivalents

The lift of the foils can be expressed as

$$F_L = C_L Q A_h V_b^2 \qquad (20)$$

in which

F_L = lift force = load (M) = anti-drift force in pounds

C_L = lift coefficient (= .4 at best L/D)

Q = dynamic constant for water = 1 slug per cubic foot

A_h = *projected* area of hydrofoils in square feet

V_b = velocity of boat in feet per second

It is evident that the immersion (projected) area (A_h) of the foils is dependent on the load as well as on V_b. In order to determine the particular V_b we seek, we must assume some constancy in the ratio of the load (M) to the area of either the airfoil or the hydrofoil. Since the hydrofoil area is the dependent variable, the constant ratio to establish is obviously

$$\frac{M}{A_a}$$

in which A_a is the area of the airfoil. Such an assumption of constancy is not completely justified on a strict geometric basis.

As the size of the boat increases, the wind's overturning moment varies as the third power of the size (airfoil area multiplied by moment arm), and the restoring moment generated by the ballast varies as the fourth power (initial hydrofoil volume multiplied by moment arm). The difference in powers permits a two-thirds power increase in airfoil area and a one-third power increase in the moment arm. The permissible advance in powers as size increases is therefore two and two-thirds for the airfoil and three for the displacement, or load.

This is not quite sufficient to support the position of constancy in the ratio of airfoil area to weight, but other changes that come with size help to make up for the deficiency. For example, taller airfoils have proportionately less root area in the turbulent air near the water surface, thereby making the lower part more efficient. Also, the higher wind speeds near the upper part give an effectively larger angle of attack, which in one way or another can be converted into a greater forward component. Both effects allow an increase in driving force with little increase in overturning moment.

The fact that hydrodynamic forces increase only with the square of the size is merely incidental to the present consideration. It does not represent a speed advantage for increasing size. The adjustment of hydrofoil area to airfoil area is strictly a function of velocity, once the ratio of airfoil area to weight is established for any size.

Critical Wind Speed

The models discussed in Part Two easily achieved a ratio of one pound of boat per square foot of airfoil. Numerically, then,

$$F_L = M = A_a \qquad (21)$$

But from Chapter IX,

$$\frac{A_a}{A_h} = 200. \qquad (22)$$

Substituting Equations (21) and (22) in Equation (20) gives

$$V_b = \sqrt{500} = 22.4 \text{ ft/sec, or } 13.4 \text{ knots}$$

This is the speed at which the lifting aerohydrofoil will sail as fast as the wind at 34 degrees off the wind.

Actually, the situation is a little better than described. We have not taken into account the displacement of the foils, which would have given equality of speeds at a slightly lower wind speed.

Velocity Proportional to Square of Wind Speed

Unquestionably, for these initial conditions the foils will be high enough in the water to justify a direct proportionality of total drag to the first power of V_b. Hence, for all other points of sailing

$$V_b = \frac{W_r^2}{1.91^2} \; \frac{\propto_a}{5} = -W_b \qquad (23)$$

Equation (23) is now the counterpart of Equation (3), derived for the non-lifting aerohydrofoil in Chapter IX. However, Equation (23) is valid only for a true wind speed of 13.4 knots. This tells us that the velocity of the boat is proportional to the square of the wind's speed, a relationship that is far more favorable than for the non-lifting aerohydrofoil, and that the preferred sailing tactic is the one that achieves high relative wind speeds rather than high airfoil angles of attack.

But the angle of attack of the hydrofoil cannot remain constant when the boat sails farther off the wind than 34 degrees. Some increase in drift must be sustained so long as greater speed is derived from an increase in side force that results in reducing the hydrofoil immersion. It is advantageous to accept the higher drift rates that accompany the higher speeds, rather than to project more foil in the water and reduce the speed on these points of sailing. To allow for this we add the following constraint:

$$\frac{\propto_h}{4} = \frac{W_r^2}{1.91^2} \qquad (24)$$

This imposes hydrofoil angles of attack proportional to the side forces on the airfoil.

One could reason that Equation (24) would also depend on foil immersion, which in turn depends on V_b, but this would not be entirely true. Increases in boat speed that come about as a result of increasing the airfoil angle of attack are not encumbered by an equivalent increase in side force, because the airfoil's resultant gains a greater forward component. Thus, at very high angles off the wind, W_r can be relatively low. Yet because airfoil angle of attack can be very high under these conditions, V_b can be high, the foil immersion small, and the drift low.

The angle θ, therefore, will be the sum of the included angle between foils and the two angles of attack ($\propto_h$ and $\propto_a$), neither of which will vary linearly: $\propto_h$ increasing and decreasing in proportion to W_r^2, and $\propto_a$ increasing slowly until maximum W_r is reached, whereupon the rate is accelerated. These conditions make it difficult to derive any simple mathematical solution. Iterative processes must now be used, wherein values of $\propto_h$ and $\propto_a$ are tested for consistency with Equations (23) and (24), and with those below, which come from trigonometric relationships.

$$\frac{W_r}{W} = \frac{\sin(\phi - \theta)}{\sin\theta}$$

$$\frac{V_b}{W} = \frac{\sin(180 - \phi)}{\sin\theta}$$

The 40-Knot Sailboat

Values computed to slide-rule accuracy are given in the table below.
The assumed wind speed is 13.4 knots.

Tabulation of Velocities

$\phi°$	$\alpha_a°$	$\alpha_h°$	$\theta°$	W_r (knots)	V_b (knots)	Upwind or downwind progress (knots)	
34	5.0	4.0	17.0	25.6	13.4	11.1	up-wind
45	5.2	5.4	18.6	29.7	18.7	13.2	
60	5.4	6.9	20.3	33.5	24.7	12.4	
75	5.8	7.8	21.6	35.1	29.2	7.6	
90	6.4	7.9	22.3	35.4	32.7	0	
105	7.5	7.0	22.5	33.8	34.7	9.0	down-wind
120	9.6	5.4	23.0	26.5	34.0	17.0	
135	17.0	4.0	29.0	19.6	26.5	18.8	
140	23.0*	3.0	34.0	15.4	23.0	17.6	

* Slotted airfoil.

Velocities for the lifting aerohydrofoil and the imaginary non-lifting version are compared in Figure 49. It is well worth noting that the more realistic craft exceeds the speed of the idealized model after $\phi = 75°$, and that it ultimately moves faster than two and a half times the wind speed, yet the price paid is only a small loss in ability to sail close to the wind. This is important mainly in narrow waters. For upwind beating, the lifting aerohydrofoil is inferior only to the idealized aerohydrofoil, not to any other boat.

A compromise more favorable to upwind sailing could be achieved by curving the hydrofoils so that their lower ends have a higher inclination than their roots. In this way the ratio of the vertical to the horizontal components could be adjusted to fit the changing conditions. More lift and more speed would be available in low winds, where sufficient immersion for resisting leeway need not be a problem, and more resistance to drift would be available in high winds, where the craft would be high enough out of the water to be in the desired speed regime.

There are other aspects of this computation worth describing. If the speed of the wind is less than 13.4 knots, the boat will be deeper in the water and will develop more drag. This means it will be slower than the wind at 34 degrees off the wind. But, on the other hand, the direction of W_r will be more favorable to closer pointing, if desired. If the wind speed is more than 13.4 knots, the boat will be higher and resist leeway less, which means it will not sail at 34 degrees off the wind when moving at the speed of the wind. However, in either case there are compensations: when sailing slower it can point up higher, and if it must point off more, it will make more speed.

For the critical wind speed under discussion, the best average progress against the wind is still made at about 45 degrees off the true wind. This angle decreases for lower winds and increases for higher winds.

The angle for best progress downwind is not

so obvious. It lies somewhere between 125 and 145 degrees off the wind, depending to a large extent on whether the airfoil can be rotated with respect to the hydrofoil, and whether slots and flaps are added to the airfoil at high angles of attack. I assumed a fixed relationship for the foils in this analysis. Although this assumption led to no unreasonable attack angles in the area of interest (unlike the non-lifting aerohydrofoil), at about 140 degrees off the wind the conditions changed sufficiently to warrant adding a slot. If the additional complexity of this device were acceptable for the design, there is no reason why it could not be employed at smaller angles, in which case the best downwind point of sailing could be close to 130 degrees off the wind.

Downwind Sailing Performance

In a very general way one can infer that a mirror image of sailing to windward exists in the downwind case. When the wind is low, more downwind progress is made by aligning the boat's direction more nearly with the wind's direction; when the wind is high, better progress is made in a direction farther from the wind's line.

Finally, for downwind sailing, an *average* speed can be obtained that is greater than the speed of the wind itself—much better, in fact, than for the imaginary aerohydrofoil. Results not quite as good are obtainable upwind. The best one can do here is to equal the wind's speed, but this still exceeds the performance of any other kind of sailboat.

The shape of the performance curve for the lifting aerohydrofoil will vary with changes in wind speed. At very low wind speeds the curve will resemble that of the non-lifting aerohydrofoil, although somewhat lower in absolute value. But since the speed of the lifting aerohydrofoil is proportional to the square of the wind's speed, the ratio of maximum speed to true wind speed should increase as the wind increases in strength. Thus, if V_b maximum is 35.4 knots for a value of W equal to 13.5[9] knots, the boat's maximum speed at higher wind speeds should be[10]

$$V_b \, max = \frac{35.4}{13.5^2} W^2$$

For a wind of 15 knots, therefore, a maximum speed of better than 43 knots should be attainable. Considering that a 15-knot wind is usually accompanied by fairly moderate waves, the surface conditions needed for such speed would not be inconsistent with the required wind.

[9] *For 35·4 read 35 and for 13·5 read 13·4*

[10] *For* V_b *max = $\frac{35·4}{13·5^2}$ W^2 read* V_b *max = $\frac{35}{13·4}$ W^2*

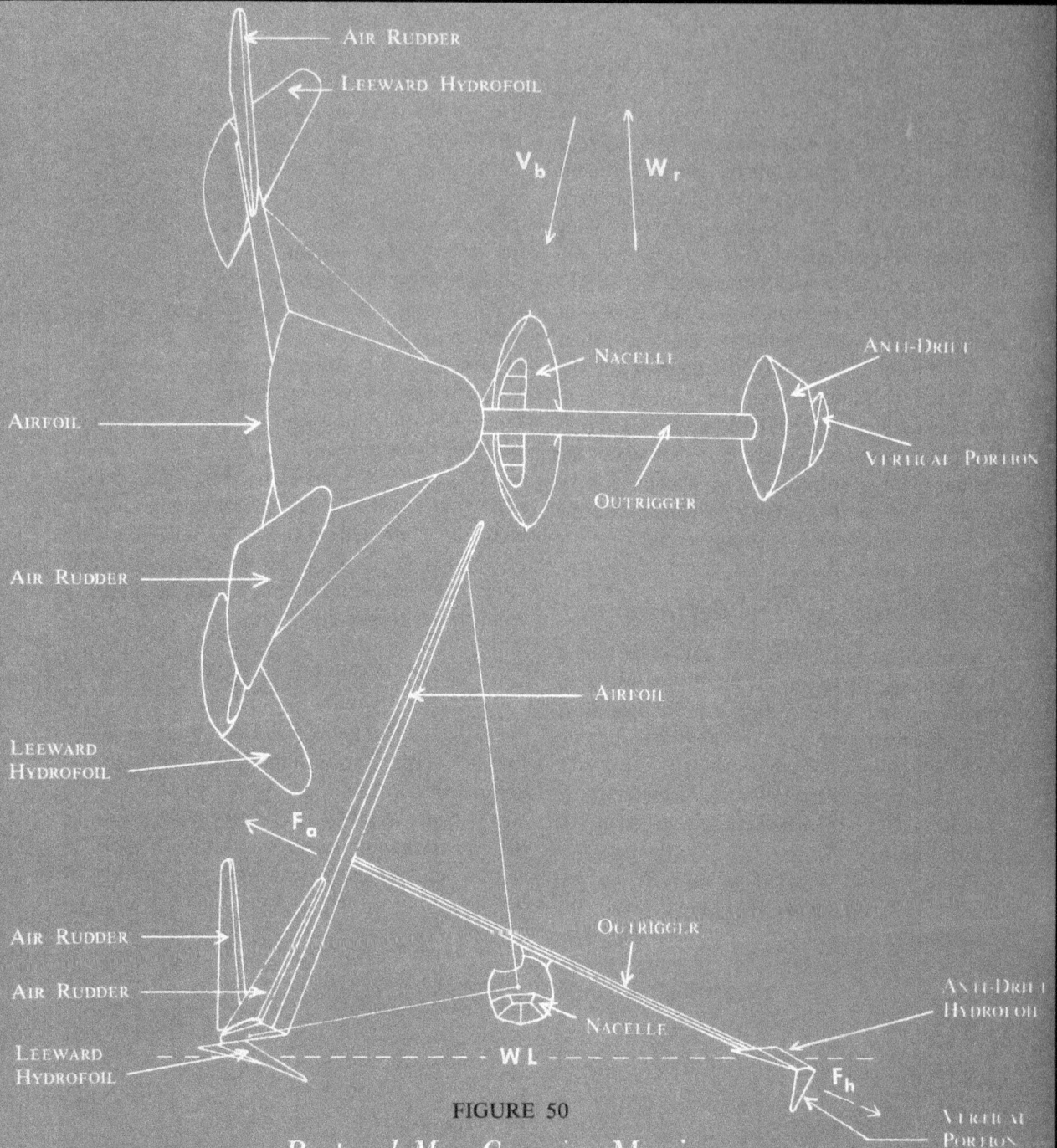

FIGURE 50

Proposed Man-Carrying Merrimac

The simplest way to give the aerohydrofoil the ability to come about is to make it reversible. In the above plan, all foils develop the required forces regardless of whether the flow is forward or backward, and all foil chords are permanently parallel except for those of the air rudders and the anti-drift foil. The lower drawing, which is a front view, indicates the balanced action of force F_a, developed by the main airfoil, and force F_h, developed by the drift-resisting hydrofoil. When these forces are equal, opposite, and in line, no overturning moments are generated.

The upper drawing is a plan view. For the setting given to the drift-resisting hydrofoil, the direction of motion is indicated by V_b. The direction of the relative wind, W_r, is shown for close-hauled sailing. Both V_b and W_r can be varied by adjustments to the after air rudder and the drift-resisting hydrofoil. To change to an opposite tack, both air rudders and the drift-resisting hydrofoil are reversed. This causes the craft to reverse direction and come about. The crew is housed in the football-shaped nacelle, which includes the boat's dinghy. In motion, the craft's waterline (WL) would be lower than is shown in the front view.

CHAPTER XII

Design of the Aerohydrofoil

The design of a practical man-carrying aero-hydrofoil is best approached in two phases. In the first phase one may think of the machine as being committed to sail on one tack at all times, like the Portuguese Man-of-War. By taking this approach, the initial design is not encumbered with requirements for coming about or for contriving, when necessary, a mirror image of the boat. The second phase encompasses the articulated joints that must be improvised to perform the inversion essential to sailing on an opposite tack.

Owing to the limitations of mechanical contrivances, the additional versatility required in the second phase will inevitably result in some performance compromises. Theoretically, a perfect conversion to a mirror image should be possible, but, practically, it cannot be obtained without great complexity. The compromise to be made, therefore, is between efficiency and simplicity.

We have already determined that the airfoil direction can be fixed with respect to the heading of the hydrofoils for all points of sailing, and that the included angle between the chords of the two kinds of foils should be about eight degrees. Only on the broadest of reaches, tantamount to running downwind, does the angle of attack on the airfoil exceed the stall angle; even then, aerodynamic stalling is not critical to the sailboat's performance.

Both analysis and experiment dictate at least three additional requirements. The first is that all the lifting hydrofoils should be inclined in the same direction. Otherwise either lift or leeway resistance is decreased on a tack. The exposed upper edges of the foils, of course, should lean to leeward.

Second it is necessary to have the buoyant force attainable with any two foils equal to at least the total weight of the boat. A little thought will show that if this requirement is not met, overturning moments can upset the boat when it is at rest before the full influence of ballasting can operate to prevent capsizing. Some further consideration will show that three foils are optimum, since this number imposes the smallest freeboard height on the foils.

Moreover, it is the minimum number required for stability and the optimum number for least stress on the boat frame.

Finally, the force resultant developed by the airfoil should be a perpendicular bisector of (but not necessarily coplanar with) a line connecting two leeward foils. Failure to achieve this relationship within rather small limits will cause one or the other of the leeward foils to submerge when the airfoil is struck by strong winds. The sensitivity becomes more critical as the distance between the leeward foils is shortened. Resorting to one leeward foil and two windward foils would be no solution to this problem; on the contrary, it would make the boat sensitive to capsizing about a line between the leeward foil and one or the other of the two windward foils.

Elimination of Capsizing Moments

The above specifications provide maximum resistance to capsizing and pitching moments, in the event such moments occur. However, they are limited in effectiveness in that their corrective reactions are not proportional to the wind's force. The complete solution requires an arrangement that generates counter-moments in proportion to the upsetting moments, or, better yet, *eliminates the source* of the moments.

To understand how the last measure can be accomplished, we shall consider the kite and reconstruct what would happen if a kite string were fastened to the floor board of a small boat. A kite furnished with suitable control surfaces could be made to fly across the wind as well as into it. A sufficiently large kite would pull the boat with it, in the direction of the string, and it would also raise the boat in the water. But only the most trivial moments would be generated on the boat.

This analogy seems to offer such a good solution to the problems of the sailboat that one may well wonder why it hasn't been used. Alas, the prospect of having a large kite lying, soggy and ruined, in the water because the wind dropped or reversed, has kept more than one brave soul from trying the experiment.

Nevertheless, this principle is the one we should employ. No moments would be generated and no ballast would be necessary if the resultant from the airfoil pointed directly away from the resultant developed by the hydrofoils.

All things considered, the requirements lead to a configuration precisely like the *Little Merrimac* (Figures 33 and 34). This is the fundamental model, and one can see how much it resembles a hybrid comprising a proa and half of a sailplane. Its components will now be examined in detail.

Hydrofoils

All foils are inclined at low angles to the horizontal in order to favor high lift, rather than leeway resistance, except for the nearly vertical extension of the windward foil. This foil *alone* serves the function of resisting drift. This arrangement puts the center of lateral resistance closely in line with the airfoil resultant. The reverse inclination of the short windward foil creates a restoring moment proportional to the wind's force and eliminates all requirements for ballast. The reverse inclination may at first seem to violate the principle of similar inclination for all the hydrofoils, as stated at the beginning of this chapter. However, it does not because it acts asymmetrically, which brings it within the intent of the postulated principle. If the entire windward foil were given reverse inclination, the performance of the boat would be impaired very seriously, as described in Chapter XI. It is necessary only that the hydrofoil that is parallel to the airfoil be large enough to develop a force equal to, opposite to, and in line with that of the airfoil. Above this hydrofoil should be placed a lifting hydrofoil bearing a suitable load. If the last requirement is not satisfied, the situation could become precarious in a seaway. If no mechanism existed for forcing the anti-drift foil to follow the water surface, this foil could leave the water when passing over the trough of a wave—an event that would leave the boat with no dynamic righting moment.

It is also essential to avoid having vertical surfaces near the waterline on the leeward side.

The sudden immersion of such surfaces in a high wave would immediately create a large overturning moment. If leeway resistance on the leeward side could be held below a critical value the hazard of exposing the windward foil would vanish, for the boat would merely slide with the wind when the drift-resisting foil lifted out of the water. This refinement in design appears to be entirely feasible.

Outward raking of the foils helps to increase longitudinal stability at high speeds, when stability becomes more critical. As the foils climb out of the water, the overall waterline is lengthened (although the foil waterlines are shortened). The reverse sweep of the forward foil does not impair its hydrodynamic properties.

Airfoil

The inclination of the airfoil causes it to lose about ten per cent of its driving force, and it diminishes the L/D by the same amount. However, by following good glider practice the net horizontal L/D of the airfoil can be held to twenty, and the L/D of the entire platform can easily be kept above ten.

The tail end of the boat provides a good position for an air rudder. Such a rudder provides better control than would a water rudder. If the hydrofoils are positioned correctly, directional stability can be achieved with a small amount of reverse rudder.

As angle of attack is increased to sail farther off the wind, the angular migration of the airfoil's resultant gives the boat more weather helm. But since the angle of attack is increased by giving the air rudder more reverse force, weather helm is essentially neutralized. Thus, trimming the air rudder sets the course with respect to the wind.

Parasite Drag

The outrigger, the fuselage, and the exposed hydrofoils together cause parasite drag. Any attempt to use the outrigger for lift generates a negative driving force, and vice versa. This surface, therefore, is best designed to give mini-

mum drag when beating to windward, where total L/D is important. If the design is correctly executed, this member is subject to little more than tensional stress, and under these conditions it can be very slender.

Payload

Since all moments are presumably balanced, no ballast is needed. The payload's position is governed only by the location of the center of buoyancy, which should lie at the centroid of the three foils. All three foils can now be loaded equally, giving the boat a higher payload capacity than is possible with a ballasted aerohydrofoil. The stability of the boat is no longer dependent on weight. Therefore, for very high-speed runs the boat should be lightly loaded.

Performance

In view of the light loading potentials, this aerohydrofoil should attain nearly the same performance as the non-lifting aerohydrofoil. Depending on the ratio of "sail" area to weight, there should exist a wind force for which the inclined hydrofoils are almost out of the water. Riding this light-fingered way, the boat becomes essentially an airfoil and a hydrofoil, as postulated theoretically. When heavily loaded, its performance would tend to duplicate the behavior of the lifting aerohydrofoil—that is, it would have reduced upwind capability and increased downwind capability, provided that the nearly vertical (windward) hydrofoil could be tilted to reduce drag.

Coming About

As we have described it up to now this boat, of course, cannot return to its starting point unless the wind takes a favorable turn. In this respect it is not much better than the earliest sailboats. The problem now is to provide the aerohydrofoil with the means for coming about. Two approaches are possible. One would be a fore-and-aft conversion, like the proa, and the other would be a left-and-right conversion in the manner of ordinary sailboats. The first is easier, although it does not escape the inevitable compromises.

Figure 50 illustrates how the fore-and-aft conversion can be accomplished without a *severe* loss in either performance or simplicity. With the exceptions of the rudder foils and the anti-drift foil, all the foils have chord lines that are permanently parallel. The convex surface of the principal airfoil is an arc that meets the opposite face, at each edge, with a radius equal to about one per cent of the chord. This sort of airfoil has fairly good efficiency with the wind flowing in either direction. The lifting hydrofoils have similar profiles. By virtue of their low inclination, these foils do not need to be reoriented for each tack. The desired angle of attack (about five degrees) is induced by the motion of the boat, which is forced, by the anti-drift foil, to follow a course about twelve degrees to leeward of the boat's heading. The anti-drift hydrofoil is trimmed for opposite tacks by swiveling it about a central hinge.

The air rudders are manipulated by cables that pass into the outrigger at its connection with the airfoil. To come about, the rudders and the anti-drift foil are swung hard over into their reverse positions. This causes the boat to stop, turn, reverse its motion, and head into the wind on the opposite tack. The forward rudder is trimmed, much like a jib, to increase the flow over the low-pressure side of the main airfoil. The after-rudder is trimmed to give the driving airfoil the desired angle of attack, which determines the heading. The rudders reverse their roles, of course, when the boat changes tack.

Location of Nacelle

Placing the payload nacelle well toward the windward side adds to the weather helm. Without it the windward foil probably would have to be shifted for each tack to compensate for the forward shifting of the airfoil's center of pressure. The illustration shows a dinghy drawn up under the nacelle. This seems to be the best location for aerodynamic reasons as well as for convenience in stowage. The dinghy can provide the floor and seats of the cockpit, and in an emergency it could be lowered, with its occupants, to the water. Another novel place for payload would be in the airfoil. A wing thick enough to hold people would not require an excessively large boat. Unlike an airplane wing, this one is nearly vertical. A person could sit or stand in a two-foot-thick wing, which is about the size associated with a 35-foot-long aerohydrofoil.

There is, of course, no fabric sail to be raised or lowered. The functions of such a sail are discharged by adjusting the inclination of the airfoil, which is done by changing the length of the two lower cables. Speed changes and adjustments for the wind's strength are also obtained in the same manner. For these functions the outrigger is hinged where it joins the airfoil. An arrangement as simple as this, however, might be suitable only for small aerohydrofoils; larger versions, in the multi-ton class, might have to use a system of fabric sail-wings for the purpose of adjusting sail area to wind strength and desired speed.

Achieving bilateral reversal is far more difficult. Every version I have designed thus far has exhibited a forbidding complexity, except the one depicted in Chapter V, which still depends on ballast for stability. I actually did try to build a bilateral version of the fundamental model described in this chapter, but with very little success. It became so complicated, with sliding joints, and so heavy, with duplicated hydrofoils, that I gave up in defeat. The proa plan is still the best, even though it does require more room to come about than an ordinary sailboat.

Three Special Problems

The aerohydrofoil is beset with three problems in a more aggravated form than is usually true for conventional sailboats. These are berthing, fouling, and striking submerged obstacles. The solution to the first problem might be special docking facilities, because it will not be easy to bring this sprawling boat up to a berth. Another possible solution would be to moor it to a float and transfer passengers and payload in a dinghy or motorboat, which could be drawn up to a nacelle as suggested earlier. Still another alternative would be to treat it like a seaplane, and either lift it out of the water with a crane or draw it up an inclined ramp on a dolly.

The problem of fouling by marine growths is quite a serious one for hydrofoils. After a week or so of immersion in most waters, the organic material accumulated on the surfaces would greatly impair the lift. The foils would require cleaning much more frequently than ordinary hulls in order to maintain good performance; that is, unless someone should come up with a finish to which nothing will adhere. The newer plastics, such as Teflon, seem to have such properties. High speed tends to diminish the problem, and in this respect the aerohydrofoil may provide its own solution if it is used often.

The pilot of an aerohydrofoil would be faced with a strange and unfamiliar mode of operation when approaching submerged barriers or shallow water. The standard reaction of a sailor to such a situation is to slow down. Yet with an aerohydrofoil, it might be prudent to speed up because of the reduction in draft that accompanies increasing speed. The contradiction between the natural impulse and the new consideration might be enough to make the hardiest "salt" neurotic. However, the risk of sinking is less than for ordinary small boats; the fact that the aerohydrofoil has three independent, widely-separated floats gives it some of the safety possessed by a boat with bulkheads.

All inventions present new problems, but somehow we learn to live with them if the advantages are great enough. To me, it seems that the aerohydrofoil is worth the effort involved in overcoming some of is inconveniences. A wind machine that, potentially, can move at almost three times the speed of the wind and withstand winds that would blow other boats flat should be permitted a few disadvantages.

New Order of Comfort

In addition to the speed and stability potentials, a few more positive aspects can be listed. A boat supported at three widely-spaced points can provide a new order of comfort to passengers located at the center of lift. Relatively large excursions can be sustained at the support points without introducing large angular or linear accelerations on the boat's occupants. Moreover, by its very nature, the support system, either in the static or dynamic phase, is a highly-damped one. This gives the aerohydrofoil riding properties more nearly akin to an automobile than a boat. It should have no sickening, long-period oscillations in pitch or roll, except perhaps for the rare circumstance when the aerohydrofoil travels over a series of uniform waves exactly in phase with its own natural frequency.

What are the size limitations? Superficially, the aerohydrofoil should be subject to the same kinds of engineering constraints that limit the size of airplanes. Actually, the aerohydrofoil escapes much of the analogy because it is initially dependent on displacement for support. Consequently, load-carrying ability increases with the cube of the scale—a very high rate, indeed. One can apportion more weight to the solution of engineering needs in large aerohydrofoils than is economical for large aircraft. For example, a *Merrimac* scaled one foot to the inch on the *Little Merrimac* plan could support a ton. Allowing for the fact that its airfoil would have to be proportionately larger than that of the *Little Merrimac* (which the physics of the platform will permit) in order to retain the same speed characteristics, its structural weight would advance to about a quarter of a ton, leaving three quarters of a ton for payload. It would be virtually impossible to scale a glider, the closest analogous platform, the same way. If we carried the argument further and raised the scale by an additional factor of three, to make a boat 100 feet long, the aerohydrofoil would begin to match in payload which can be carried profitably by a *powered* air transport of the same general dimensions—about 25 tons.

The Potentials

Although there are practical reasons for seeking high-speed sailboats, I discuss them without any desire to add more "practical necessities" to an already overburdened culture. Our lives now are filled to overflowing with the chores of selecting, possessing, cleaning, maintaining, repairing, disposing of, and avoiding payment for all the goods designed to make living simpler and easier. I would be perfectly willing to stress the aesthetic and recreational reasons for increasing the speed of sailboats, and advance no others. But I must bow to the cause of completeness. There are military, commercial, and scientific uses for high-speed sailing ships, and these uses could be important in the future.

The sea is a mysterious and often cruel environment, perhaps full of more unknowns than any other environment on this planet. Yet man, who always strives for superlatives, has surpassed nature with the new and fearful attributes he has bestowed on the sea. It now hides the stealthiest and most destructive machine ever conceived by the mind of man—the submarine. These days we hear much about the military threats presented by space machines, just as we heard about air power before and during World War II. It is all very impressive until we learn that air raids were less critical to the outcome of the war than were submarine attacks on shipping. The only exception was the nuclear bombing of Japan, and even acts of destruction like this can now be performed by the submarine, with far more effectiveness and far less vulnerability.

Certainly when one is at sea and tries to look down into the opaque vastness of inner space, outer space seems the lesser threat. Only in the rarest of waters can anything more definite than patches of light and darkness be discerned visually beyond a hundred feet. With some very low radio frequencies the penetration is a little better, but hardly enough to make a significant difference in detecting submarines—and even then the wave lengths are

126

too great to suit the purpose. The whole spectrum of radiation has been carefully examined for a wave length that might provide a "hole," but the only practical method of detection is still the old one, sound.

Sea animals have spent many, many more years than man in the same quest, without finding anything better than sound for long-range communication and detection. Of course one cannot argue that this settles the question forever, but it does diminish any hope of a breakthrough. Man's best prospect of hunting down the submarine today rests on his ability to imitate the fishes and sea mammals, which perform wonders with sound alone.

Silence—A Tactical Advantage

Picking up information about the presence of a submarine requires careful listening, and, as every competent conversationalist knows, good listening requires silence. Of all surface vessels, the sailing ship has the highest potential for silence either at rest or in motion. It creates no internal motor noises or external propeller noises to interfere with and confuse any signals from the surrounding sea. Practical attempts to realize this potential during wartime have not been entirely successful because of the squeaking blocks and clanging chains carried by large sailing ships. However, modern materials could easily eliminate such nuisances in a sailboat designed specifically for hunting submarines.

If it were truly as quiet as it can be made to be, the sailing ship would have a great passive listening advantage over the submarine, and under certain conditions the submarine's propulsion noises could be detected by the sailing ship long before it, in turn, could be heard by the submarine. This is a military advantage of the first importance; no end of mischief could be perpetrated on a target that did not know it had been detected. A submarine would need to extend its periscope or use echo-location in order to gain awareness of this enemy, a tactic that submarines are unlikely to use in the future because it increases their own vulnerability to detection.

A quiet sailing ship, capable of moving upwind and downwind as fast as the wind, would have some additional advantages over other ships in combing windy expanses for submarines. It would not be tied to fueling stations or require much in the way of underway replenishment, a freedom of some military importance. And in the North Atlantic Ocean, at least, the sustained average speed of an aerohydrofoil should be considerably higher than that of any motorized craft of the same displacement. (For the rare calms that exist over this part of the ocean the ideal auxiliary propulsion plant would be a motor-driven air propeller. There would be no head-wind problem for the air propeller in a calm. Noises acoustically coupled to the water from the usual cavitating water propeller could be avoided with an air propeller, as could the maintenance problems associated with submerged rotating parts.)

Many other natural attributes of the sailboat hold valuable potentials for antisubmarine warfare. For example, sailboats can be made not only nonmagnetic, but entirely nonmetallic, thus diminishing their vulnerability to mines and improving the sailboat's own ability to detect magnetic and electric anomalies created by either mines or submarines.

And so the ancient struggle between the Portuguese Man-of-War and the fishes may yet make a full turn in modern times. The sailing ship may yet engage the submarine and come out the winner.

An Elusive Target

In time of war, sailing ships that could exceed the speed of the wind, as iceboats do, would have a fair chance of outrunning submarines along the great-circle shipping routes in the North Atlantic and North Pacific oceans where the wind velocity is 10 to 25 knots consistently. In addition to being difficult to catch, such ships would be difficult to kill. The classic tactic of firing an array of torpedoes against a long, highly vulnerable waterline would not be available to the submariner. Three shallow, widely spaced hydrofoils would present a far more elusive target and, as a matter of fact, an entirely new problem. The probability is also good that when mass-produced, such vessels would cost less than the torpedoes required to sink them, and they would be able to carry considerable loads. The spectacle of a submarine racing after a school of relatively small sailing ships, no one of them worth a torpedo and only a few of them apt to be hit, would be enough to discourage the most optimistic submariner.

But superior speed and low initial cost would not, by themselves, permit the sailing ship to re-enter the field of commercial cargo movement, even in time of war. There must also be a substantial reduction in crew size— otherwise, the steamship's main economic advantage over the sailing ship would continue to be the deciding factor, particularly in times of high labor costs. I willingly leave this problem to others with more optimism.

For oceanographic research requiring vessels with nonmagnetic properties, silent operation, long range, shallow draft, and high speed, the aerohydrofoil might be uniquely useful, particularly if it were a fairly stable platform both at rest and in motion, as some of the models have been. What could be more admirable, or more conducive to accurate scientific research, than being transported on a platform that is in complete harmony with the environment being studied?

Best of All Potentials

The enjoyment of life is also a practical matter. For this purpose the pleasures of sailing are obvious and need no elaboration. A man-carrying aerohydrofoil would gladden the heart of any true seaman, whether he sails for the contest with nature or to race with other boats. Should the fates willingly lend a hand, a standardized, low-cost version could be available to him at the time the sequel to this volume is published—a sequel that, hopefully, will describe the final phases of full-scale aerohydrofoil development; a sequel that I pray will not be long delayed.

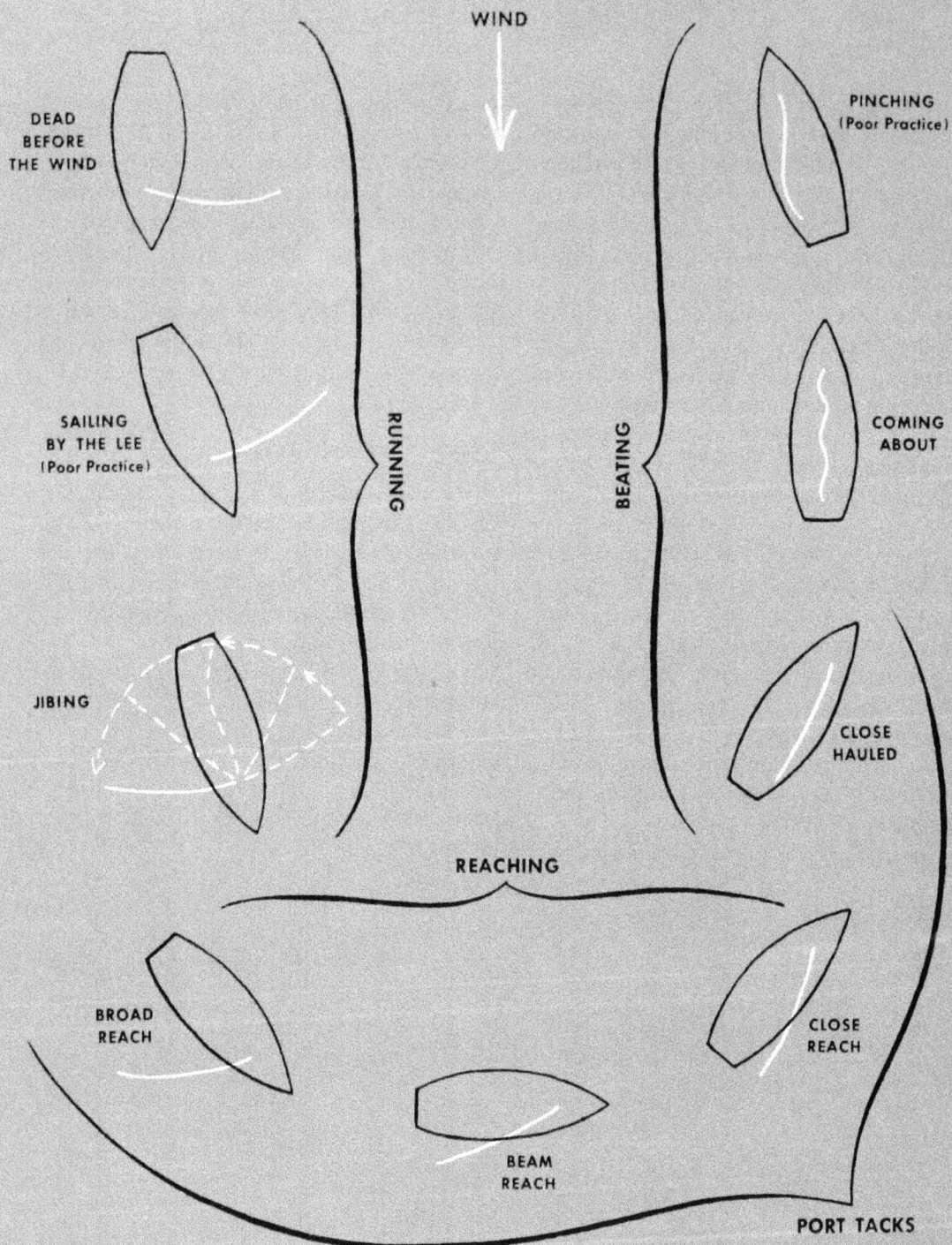

WIND

DEAD BEFORE THE WIND

PINCHING (Poor Practice)

SAILING BY THE LEE (Poor Practice)

RUNNING

BEATING

COMING ABOUT

JIBING

CLOSE HAULED

REACHING

BROAD REACH

CLOSE REACH

BEAM REACH

PORT TACKS

FIGURE 51

Points of Sailing

The sailing points designated above have meaning only with respect to the true wind. If the apparent wind experienced on the boat is used as a reference, the designations will be in error. For example, a fast boat moving around the points of sailing identified as "port tacks" would feel little change in the direction of the apparent wind—not over 30 degrees—yet the boat's orientation to the true wind might have changed by more than 90 degrees, from downwind to upwind. All points of sailing are illustrated in the sketch except the precise point of change from running obliquely downwind to reaching downwind. One definition, not widely accepted, is that a sailboat is "running" if its sail catches the wind like a cup and "reaching" if its sail receives the wind flow like an airfoil.

Appendix: Simple Theory of Sailing

Running before the wind (Figure 51) is so obvious a maneuver that it needs no elaboration. It is equivalent to drifting. The drift rate may be increased by either or both of two methods: by increasing the sail area exposed to the wind, as the clippers did, or by making the sail more of a cup, as modern boats do with spinnakers and ballooners.

The speed of a sailboat downwind is always limited to something less than the speed of the wind, because the force on the sail *decreases* as the boat speeds up. As the boat partakes of some of the wind's speed, the relative wind on the sail diminishes by the same amount. A sailboat could reach the speed of the wind on a downwind sailing only if the resistance in the water were zero, which is impossible. Running downwind is not the fastest point of sailing for many modern boats, although it was the best for square-riggers.

A boat designed only for sailing downwind can also sail at an angle to the wind, but the boat will drift off course. Sailing at an angle to the wind with a minimum of drift requires the addition of a relatively flat projection that extends below the waterline in line with the axis of the boat. Such a projection—a keel or similar structure—offers resistance to side-drift, or leeway. It also makes the boat capable of sailing against the wind. This is so because the force developed by the sail is nearly perpendicular to the plane of the sail for almost all wind directions. Therefore, by arranging a suitable angle between the sail and the keel, forward motion can be generated. To clarify this and other statements, two theoretical sailing terms are introduced:

1) *Center of Effort* (*C.E.*) If all the force of the wind collected by the sails were applied to one point on the rigging, so that it gave the same motion to the boat as the sails did, this point would be the C.E.

2) *Center of Lateral Resistance* (*C.L.R.*) If all the water force acting on the boat to resist leeway were concentrated on one under-

130

water point, so that it duplicated the reactions of the hull and keel, this point would be the C.L.R.

From Figure 52 it is apparent that an important reaction of the boat to the spacing of the centers is to change its direction of motion. If the centers are in the same vertical line, the boat is directionally balanced but will steer uncertainly. If the C.E. is aft of the C.L.R., the boat will tend to turn into the wind, in which case it is said to have "weather helm." Most boats are purposely designed with weather helm for safety. If the rudder were lost, or the helmsman incapacitated, a boat so designed would head into the wind and luff. The achievement of weather helm is not as simple as this dissertation might imply. A heeling boat has more complicated forces originating in the sail and hull, and these must also be taken into account.

The second important reaction to the spacing of the centers is in the magnitude of the overturning, or capsizing, moment. The greater the vertical spacing of the centers, the greater is the overturning moment created by the wind. Of course the righting moment, or the torque that keeps the boat erect, also depends on the depth of the keel, if the keel is weighted. The centers can be brought closer together by making the sail and the keel stubbier, or by bringing the sail and keel closer to the waterline. The first kind of adjustment is easier, but it reduces the upwind sailing efficiency of the boat. Practical requirements usually interfere with attempts to make the second kind of adjustment. The foot of the sail is generally raised above the heads of the boat's occupants to increase visibility and to avoid skull injuries which could be sustained in the excitement of coming about. As a consequence, the C.E. is elevated 3 to 5 feet higher than it would have to be if sails were transparent and sailors more nimble. Similar problems keep the keel or centerboard from being brought closer to the waterline. Only by using a leeboard at the

side of the hull or by eliminating the hull, as in the aerohydrofoil, can the C.L.R. be raised significantly.

The interaction of the two centers is of importance in downwind sailing too, although under downwind conditions the forces corresponding to the centers are perpendicular, and no heeling moment arises from the difference in height of the centers. Pitching moments are affected, however, and the C.E. does influence the lateral turning moment and frequently puts a heavier burden on the rudder than does sailing on a tack. These various requirements result in compromising a sailboat design to the extent that it never quite achieves the ultimate performance that would be possible otherwise in any one direction.

When the centers are aligned properly, as shown in Figure 52, part of the force developed by the sail can be applied to the forward motion of the boat while the remainder is resisted by the boat's underwater surfaces. The interaction of these forces can be demonstrated vividly by a simple experiment. Prepare a bar of soap by shaving one side so that it is not quite parallel to the opposite side. One side can now be considered in line with the keel, and the other with the sail. Apply the pressure of moistened thumb and forefinger to the sides of the bar. If your fingers are out of line, the bar will twist and fall out of your hand. You have not applied the forces to the proper C.E. and C.L.R. and the "boat" has changed direction. But with thumb and forefinger exactly opposing each other, the soap will squirt forward. If you think of your thumb as the force resisting leeway, and of your forefinger as the force on the sail, the process of sailing upwind will become clear. You may also have observed that the soap moves forward faster than your fingers can come together. This demonstrates how it is possible for a boat to sail faster than the wind.

It is of course impossible to sail directly into the wind. Therefore, to make progress upwind,

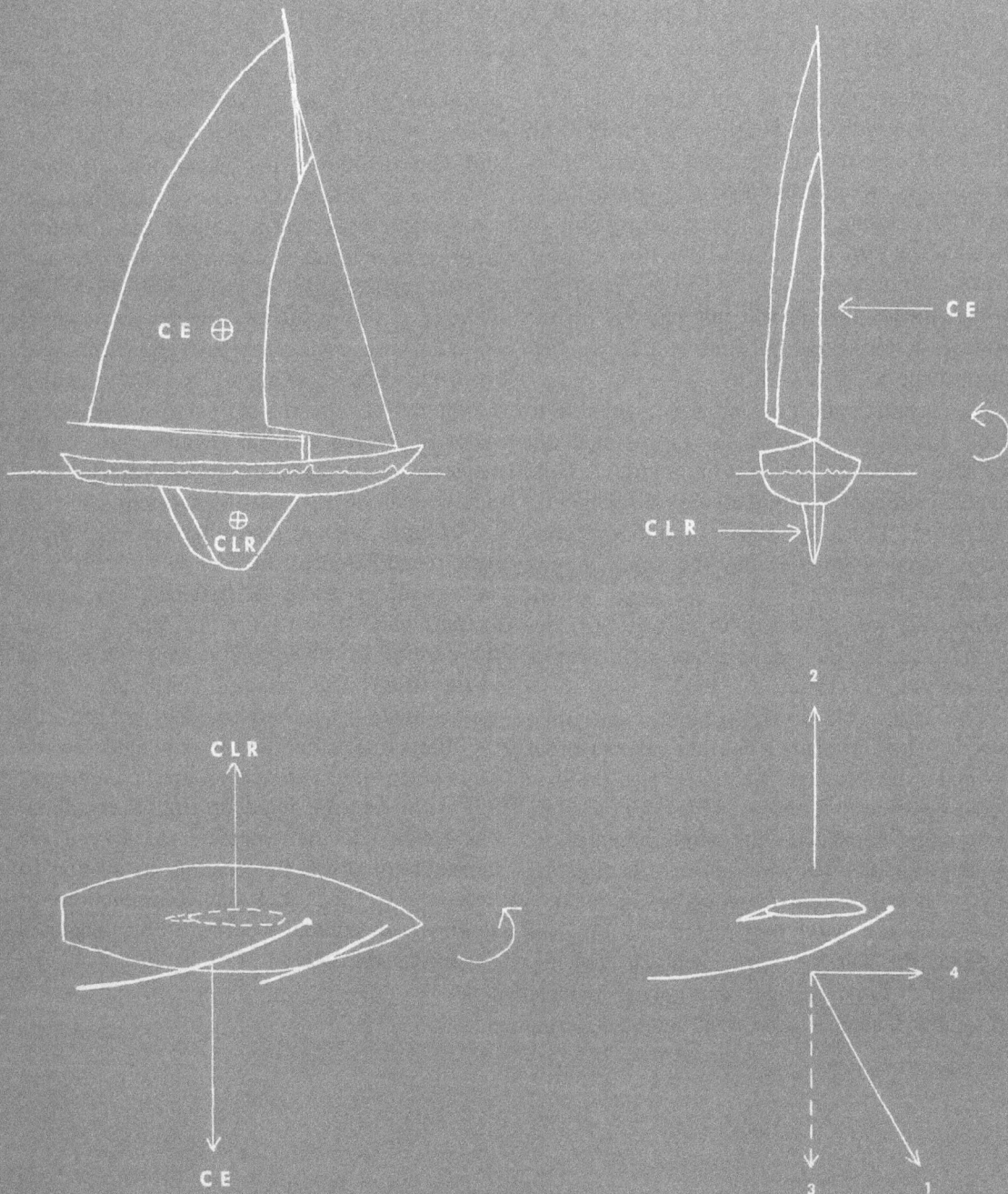

FIGURE 52

Forces on the Sailboat

The side view shows the location of the Center of Effort (C.E.) and the Center of Lateral Resistance (C.L.R.). The front and plan views show how the two centers affect overturning and directional turning. In the diagram at lower right, the force developed by keel and rudder (2) cancels the side component (3) of the force (1) generated by the sail, leaving only the forward driving component (4) uncancelled.

it is necessary to beat to windward (Fig. 51). In this procedure the boat is sailed as close to the wind as is practicable, and at periodic intervals it is brought about on an opposite tack, each time losing as little speed as possible in making the turn. Should the boat lose headway when coming about, it is said to be "in irons," which means that it will not respond to the rudder. A very good boat can average as well against the wind as with it.

In an extremely fast boat it is profitable to tack downwind as well. As was pointed out earlier, there are inherent limitations to the speed a sailboat can make going directly with the wind. There are fewer limitations, however, when a sailboat has the wind coming partly from the side. Thus, whenever a boat offers very low resistance to motion (as does an iceboat), it can make faster time *beating* downwind.

Changing tacks downwind, particularly in a strong wind, presents some hazards. In order to make the best time, the sails must be jibed from one side to the other, which is equivalent to coming about with the wind from behind. In the transition the wind reverses its direction over the sail. If the operation is not planned carefully, the wind may catch the opposite side of the sail too soon, swing it violently through a large arc, and carry away the rigging. For a small, unballasted boat in a strong

wind, the safest thing to do is turn the boat upwind, come about in the orthodox fashion, and then turn downwind on the opposite tack.

Distinctly different kinds of thinking are required for operating a sailboat and a motorboat. A dead calm is a nuisance for one and a boon to the other. A strong wind heralding a storm may be a source of worry to a motorboater whose engine has been drenched by a wave or whose fuel supply is low; the same conditions can be reassuring to a competent sailor in a snug sailboat because the storm's warning winds carry the power to take him safely home, if he prudently shortens sail.

A classic sailboat puzzle will serve to highlight the differences. "Under what conditions can a sailboat make the best headway downstream: when the wind moves precisely with the current, or when it is at rest with respect to the land?" The answer is surprising. Faster way is made when the wind does *not* move in the direction of the current. The sails feel no force when wind and current are in alignment; but with the true wind at rest, the sailboat has headwinds that enable it to *exceed* the current's progress by beating to windward downstream. Such are the contradictions of sailboats. A sailboat would do best under the conditions that would most hamper a motorboat.

Bibliography

ABBOT, VAN DOENHOFF: *Theory of Wing Sections, Including A Summary of Airfoil Data.* Dover Pub., Inc.; New York, 1960.

BAUDIN, LOUIS: *Daily Life in Peru.* The Macmillan Co.; New York, 1962.

BUCHSBAUM, RALPH AND MILNE, LORUS J.: *The Lower Animals—Living Invertebrates of the World.* Doubleday & Co.; Garden City, N. Y., 1960.

BURNETT, CONSTANCE BUEL: *Let the Best Boat Win.* Houghton, Mifflin Co.; Boston, 1957.

CASSON, LIONEL: *The Ancient Mariners.* The Macmillan Co.; New York, 1959.

CASSON, LIONEL: "Fore and Aft Sails in the Ancient World," *Mariner's Mirror.* Cambridge University Press; Jan. 1954, Jan. 1961.

CHAPELLE, HOWARD I.: *The History of American Sailing Ships.* W. W. Norton & Co.; New York, 1935.

CURRY, MANFRED: *Yacht Racing—Aerodynamics of Sails and Racing Tactics.* Chas. Scribners & Sons; New York, 1948.

DOMMASCH, SHERBY & CONNALLY: *Airplane Aerodynamics.* Pitman Pub. Corp.; New York, 1961.

FOLKARD, HENRY COLEMAN: *The Sailing Boat.* Edward Stanford; London, 1901.

HADDON, A. C. AND HORNELL, JAMES: *Canoes of Oceania, Vol. I.* Bernice P. Bishop Museum; Honolulu, 1936.

HEYERDAHL, THOR: *Kon-Tiki;* Rand, McNally; New York, 1950.

HORNELL, JAMES: *South American Balsas, The Problem of Their Origin.* Cambridge University Press; 1942.

HYMAN, LIBBIE HENRIETTA: *The Invertebrates—Protozoa Through Ctenophora.* McGraw-Hill Book Co.; New York, 1940.

LANDSTRÖM, BJÖRN: *The Ship—An Illustrated History.* Doubleday & Co.; New York, 1940.

LANE, CHARLES E.: "The Portuguese Man-of-War." *Scientific American,* Vol. 202, No. 3 (1960), pp. 158-168.

McINTYRE, MALCOLM: "The Sailplane." *Yachting,* Vol. 55, No. 2 (1934), pp. 62-68.

McINTYRE, MALCOLM AND T. A.: "The Sailplane—A New Type of Sailboat." *Yachting,* Vol. 28, No. 5 (1920), pp. 248-250.

NAESETH, RODGER L.: "An Exploratory Study of a Parawing as a High-lift Device or Aircraft." Langley Research Center, Langley Field, Va., *Technical Note—D-629,* National Aeronautics and Space Administration, Washington, D.C., Nov. 1960.

ROGALLO, FRANCIS M., LOWRY, J. G., CROOM, D. R., AND TAYLOR, R. T.: "Preliminary Investigation of a Paraglider." Langley Research Center, Langley Field, Va., *Technical Note—D-443*, National Aeronautics and Space Administration, Washington, D.C., Aug. 1960.

SAVILOV, A. E.: "The Siphonophorae Velella and Physalia in the Pleuston of the Pacific." *10th Pacific Science Congress*, Honolulu, Aug. 1961.

ROUSE, HUNTER: *Elementary Mechanics of Fluids.* John Wiley & Sons; New York, 1960.

SWEENEY, T. E.: *Exploratory Sailwing Research at Princeton.* Princeton Univ.; Dec., 1961.

VON HAGEN, VICTOR W.: *Realm of the Incas.* New American Library of World Literature; New York, July, 1960.

WALLACE, BILL: *Sailing.* Golden Press; New York, 1960.

WOODCOCK, A. H.: "Dimorphism in the Portuguese Man-of-War." *Nature*, Vol. 178 (1956), pp. 253-255.

134

---

* Footnote to Velella (see page 10).

As this book went to press the following information was received from G. E. MacGinitie, Professor Emeritus, California Institute of Technology, Consultant in Marine Biology, Naval Missile Center, Point Mugu, California.

Velella lata is a true by-the-wind sailor. By turning down a portion or even an entire side of the float so that it cuts vertically into the water, *Velella* creates an effective keel.

The arc-shaped sail of *Velella* is set at an angle of about 45 degrees to the long axis of the body. Thus the animal makes much faster progress than if the sail were at right angles to the long axis of the body in which case it would have to move with the wind directly astern.

The increase in speed resulting from the angle of the sail and the dropping of the keel, brings *Velella* in contact with much more zooplankton upon which it feeds. This discovery indicates that *Velella*'s sailing habits are purposeful and, in fact, beneficial to the animal.

Index

(PAGE REFERENCES TO ILLUSTRATIONS OR MAPS ARE IN ITALICS.)

136

138